Student Soluti

Introduction to Statistics and Data Analysis

THIRD EDITION

Roxy Peck
California Polytechnic State University, San Luis Obispo

Chris Olsen
George Washington High School, Cedar Rapids, IA

Jay Devore
California Polytechnic State University, San Luis Obispo

Statistics
The Exploration and Analysis of Data

SIXTH EDITION

Roxy Peck
California Polytechnic State University, San Luis Obispo

Jay Devore
California Polytechnic State University, San Luis Obispo

Prepared by

Mary Mortlock
California Polytechnic State University, San Luis Obispo

THOMSON

BROOKS/COLE

Australia • Brazil • Canada • Mexico • Singapore • Spain • United Kingdom • United States

Printed in the United States of America
1 2 3 4 5 6 7   08 07 06 05 04

Printer: Thomson/West

ISBN-10: 0-495-11876-1
ISBN-13: 978-0-495-11876-3

Thomson Higher Education
10 Davis Drive
Belmont, CA 94002-3098
USA

For more information about our products,
contact us at:
Thomson Learning Academic Resource Center
1-800-423-0563

For permission to use material from this text or
product, submit a request online at
http://www.thomsonrights.com.
Any additional questions about permissions can be
submitted by email to **thomsonrights@thomson.com.**

Table of Contents – Students Edition

Chapter 1

1.1   Descriptive statistics is made up of those methods whose purpose is to organize and summarize a data set. Inferential statistics refers to those procedures or techniques whose purpose is to generalize or make an inference about the population based on the information in the sample.

1.3   They are from a sample. Only *some* travelers were polled and the results led to *estimates* for the population of interest, not the exact percentage.

1.5   The population of interest is the entire student body (the 15,000 students). The sample consists of the 200 students interviewed.

1.7   The population consists of all single-family homes in Northridge. The sample consists of the 100 homes selected for inspection.

1.9   The population consists of all 5000 bricks in the lot. The sample consists of the 100 bricks selected for inspection.

Exercise 1.11 – 1.25
1.11   **a.** categorical
         b. categorical
         c. numerical (discrete)
         d. numerical (continuous)
         e. categorical (each zip code identifies a geographical region)
         f. numerical (continuous)

1.13   **a.** continuous
         b. continuous
         c. continuous
         d. discrete

1.15   **a.** Gender, Brand of Motorcycle and Telephone area code.
         b. Number of previous motorcycles owned.
         c. Bar chart.
         d. Dot plot.
         Most summer movies have box office sales of between $50 million and $152 million. There is a small cluster of 3 films that have sales of about $200 million. The two top box office totals for the summer of 2002 were significantly higher: Star Wars, Episode II at $300.1 million and Spider-Man at 403.7million.

1.17

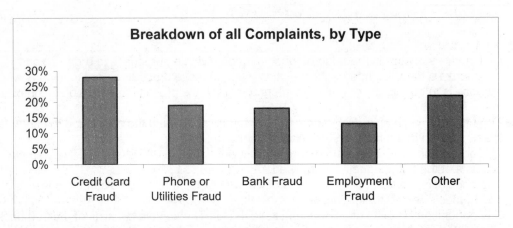

The most common single type of identity theft is credit card fraud with 28% of the total complaints, followed by phone or utilities and bank fraud, both just under 20%. Employment Fraud is less at 13%.

1.19

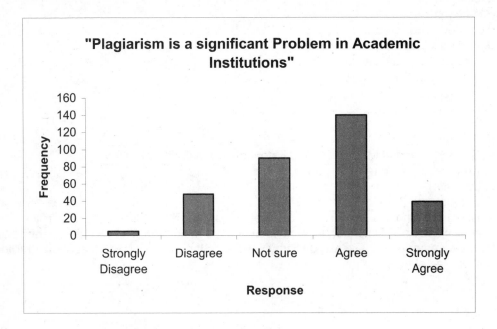

1.21

Comparison of Athletes and Overall Graduation Rates between Sports

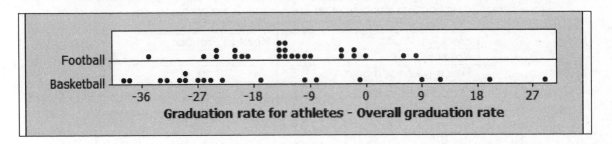

For both sports, there are few Universities where the overall graduation rate is lower than that of the scholarship athletes. At the schools where athletes do better than the overall rate, there are more schools where the basketball players do well, however, there are some schools where the basketball players graduate at a much poorer rate than overall. Overall, the athletes of both sports, graduate about 10% lower than overall, but the range for football players is lower (35% lower to 8% higher) than for basketball players (39% lower to 29% higher).

1.23   a.

Grade	Frequency	Relative Frequency
A+	11	0.306
A	10	0.278
B	3	0.083
C	4	0.111
D	4	0.111
F	4	0.111
Total	36	1.0

Water quality ratings of California Beaches

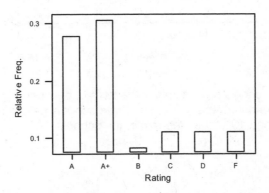

Assuming that an A+ means little risk of getting sick, most beaches in California seem quite safe. Two thirds of the beaches are rated a B or higher.

b. No, a dotplot would not be appropriate. "Rating" is categorical data and a dotplot is used for small numerical data sets.

1.25 The relative frequencies must sum to 1, so since .40+.22+.07 = .69, it must be that 31% of those surveyed replied that sleepiness on the job was not a problem.

Sleepy at Work?	Relative Frequency
Not at all	0.31
Few days each month	0.40
Few days each week	0.22
Daily Occurrence	0.07

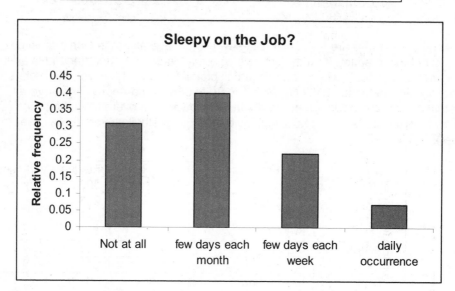

Exercise 1.27 – 1.31
1.27

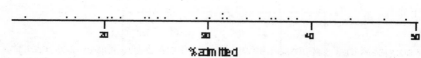

Most U.S. business schools have acceptance rates of between 16% and 38%. One school has a slightly lower rate than this (12%) and three schools have a much higher acceptance rate (between 43% and 49%) than the rest of the schools.

1.29

Sport	Frequency	Rel. Freq.
Touch Football (TF)	38	0.226
Soccer (SO)	24	0.143
Basketball (BK)	19	0.113
Baseball/Softball (BA)	11	0.065
Jogging/Running (JR)	11	0.065
Bicycling (BI)	11	0.065
Volleyball (VO)	7	0.042
Others (OT)	47	0.280
	168	0.999

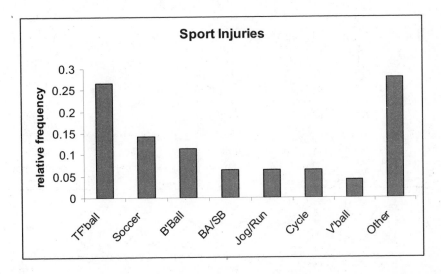

1.31 The display suggests that a representative value is around .93. The 20 observations are quite spread out around this value. There is a gap that separates the 3 smallest and the 3 largest cadence values from the rest of the data.

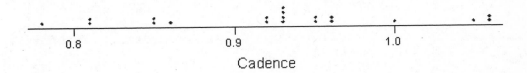

Chapter 2

Exercises 2.1 – 2.9

2.1   **a.** It is an observational study, since no treatment was imposed.
b. No. Cause-and-effect conclusions cannot be made from an observational study.

2.3   Better educated, wealthier, and more active people tend to have better health care and look after themselves better which may give alternative reasons for reduced risk of heart disease.

2.5   This is an observational study – the diabetes (the treatment) was not imposed by the experimenter and so the results were simple being observed. No cause-and-effect conclusion can be made on the basis on an observational study.

2.7   **a.** If the definition of affluent Americans was having a household income of over $75,000 and that the sample was a simple random sample.
b. No. This sample is not representative of all Americans, since only affluent Americans were included.

2.9   It is possible that other confounding variables may be affecting the study's conclusion. It could be that men who eat more cruciferous vegetables are also making a conscious choice about eating healthier foods. A definitive causal connection cannot be made based on an observational study alone.

Exercises 2.11 – 2.29

2.11   The names appear on the petition in a list ordered from first to last. Using a computer random number generator, select 30 random numbers between 1 and 500. Signatures corresponding to the 30 selected numbers constitute the random sample.

2.13   Stratified sampling would be worthwhile if, within the resulting strata, the elements are more homogeneous than the population as a whole.
a. As one's class standing increases, the courses become more technical and therefore the books become more costly. Also, there might be fewer used books available. Therefore, the amount spent by freshmen might be more homogeneous than the population as a whole. The same statement would hold for sophomores, juniors, seniors, and graduate students. Therefore, stratifying would be worthwhile.
b. The cost of books is definitely dependent on the field of study. The cost for engineering students would be more homogeneous than the general college population. A similar statement could be made for other majors. Therefore, stratifying would be worthwhile.
c. There is no reason to believe that the amount spent on books is connected to the first letter in the last name of a student. Therefore, it is doubtful that stratifying would be worthwhile.

2.15   1.   Cluster sampling
2.   Stratified random sampling
3.   Convenience sampling
4.   Simple random sample
5.   Systematic sampling

2.17   Using a convenience sample may introduce selection bias. Suppose I wanted to know the average heights of students in all Stats classes, and used the students in my 10:00 class as my subjects. It may be that all the basketball players (that are generally tall) are in my 10:00 class because they have an early practice. Not only would my non-basketball 8:00 students be excluded from being part of the sample, but resulting average height would over-estimate the average height of my students. The sample must be representative of the population.

2.19   **a.** Using a random number generator, create 3 digit numbers. The 3 digits represent the page number, and the number of correctly defined words can be counted. Suppose we want to choose 10 pages in the book. Repeat this process until 10 page numbers are selected and count the number of "words" on each page.

   b. As the number of words on a page may be influenced by which topic being discussed – and therefore which chapter, it makes sense to stratify by chapter. Suppose we want to sample 3 pages in each chapter. For the first chapter and using a random number table, generate a 2 digit number. Use this to select a page within the chapter. Repeat this process twice more until 3 pages are chosen. Find these page numbers in the chapter and count the number of defined words on that page. Do this for each chapter until 3 pages from each chapter are chosen.

   c. Choose the 10th page and then every 15th page after that.

   d. Randomly choose one Chapter and count the words on each page in that chapter.

   e.& f. Answers will vary

2.21   In theory, this is a good example of stratified random sample. Randomization is used in selecting the households in California. Another random technique is used to randomize the available subjects in each family. However, there are several possible sources of bias. It will eliminate households without a phone. It may be that the person who has recently had a birthday may be busy or too grumpy to talk to someone and some-one else (who may not be of voting age) responds to the questions. Someone may lie and give responses that are not true.

2.23   This was a small sample of people who chose to attend the workshop and were interested in dealing with the city's financial aid deficit. They are probably not representative of the population of interest.

2.25   No, it is a volunteer sample with no random selection. There is no reason to think the sample is representative of U.S. adults. Volunteers may have had a reason to be worried about mercury contamination.

2.27   It is possible that different subpopulations used the website and the telephone. The different ways of responding may have attracted different types of people, and it is possible that neither group is representative of the town's residents.

2.29   The best method to use would be to use a simple random sample where every different sample has an equal chance of being selected. For example, if each student in the school is numbered, you could use a random number generator to select a random sample of them. you could then find them, ask their opinion and then assume their views are representative of the whole schools.

Exercises 2.31 – 2.39

2.31   The response variable is the amount of time it takes a car to exit the parking space. Extraneous factors that may affect the response variable include: location of the parking spaces, time of day, and normal driving behaviors of individuals. The factor of interest is whether or not another car was waiting for the parking space.

   One possible design for this study might be to select a frequently used parking space in a parking lot. Station a data collector at that parking space for given time period (say 2pm-5pm) every day for several days. The data collector will record the amount of time the driver uses to exit the parking space and whether another car was waiting for that space. The data collector will also note any other factors that might affect the amount of time the driver uses to exit such as buckling children into their car seats. Additionally, this same setup should be replicated at several different locations.

2.33   **a.** Randomly select a reasonably large sample of students from your school and also choose 2 comparable IQ tests. Call the tests Test 1 and Test 2. Randomly divide your sample into 2 groups and administer Test 1 to the first group and Test 2 to the second group. Record the IQ scores for each student. Now gather all the students in the same room and let them listen to a Mozart piano sonata. After playing the sonata, ask the students to return to their groups. This time administer Test 2 to the first group and Test 1 to the second group. Again, record the IQ scores for each student.

    b. This design includes direct control for several extraneous variables. Each student in the sample will take an IQ test before and after listening to the Mozart sonata. The IQ tests are different but comparable each time. Two IQ tests were used to eliminate the possibility that retaking the same test may make the test easier the second time and thus higher scores because of retaking the same test may be confounded with the effects of the piano sonata. The IQ tests were given such that one group started with Test 1 and the other group started with Test 2. After listening to the sonata, the groups retook the IQ tests, this time taking the test they had not previously taken. This design was chosen to eliminate the possibility that one test is more difficult than the other. All the students in the sample listened to the sonata in the same room, at the same time and under the same conditions, therefore the factor 'listening conditions' can be ruled out as a confounding factor.

    c. This design has 'student' as a blocking factor. Each student's IQ test score was recorded before and after listening to the Mozart sonata. This design should 'block out' the effects of differences in IQ scores across students.

    d. By randomly selecting the sample and randomly placing students into 2 groups, we expect to create 'equivalent' experimental groups and minimize biases due to unknown, uncontrolled factors.

2.35   **a.** Blocking. Homogeneous groups were formed by height.

    b. Direct control.The researchers only used right handed people.

2.37   So many other factors could have contributed to the difference in the pregnancy rate. The only way that the difference between the two groups could have been attributed to the program was if the 2 groups were originally formed by dividing all the students completely randomly. This would minimize the effects of any other factors leaving the program the only <u>big</u> difference between the 2 groups.

2.39   Yes, blocking on gender is useful for this study because 'Rate of Talk' is likely to be different for males and females.

Exercises 2.41 – 2.49

2.41   It is necessary to divide into groups that are similar by a certain factor and eliminate any differences in responses. For examples, if you suspect the reaction to a drug may be different between the genders, it would be better to block by gender first before randomly dividing into the treatment groups and the place groups.

2.43   **a.** It is important to have comparable groups. It may be that people who have a mellow personality (and includes the physical characteristics of positive attitude, lower blood pressure, and using fewer laxatives) enjoy art and so if they were allowed to choose groups they would all go and discuss works of art. The two groups should be equivalent.

b. A control is need for comparison. It may be the social aspect of meeting other people that makes the physical changes so a control provides a basis for comparison that allows the researcher to determine if the discussion of art is really what is affecting the women.

2.45 If either the dog handlers or the experimenters knew which patients had cancer, they might give physical clues (consciously or unconsciously) or simply be rooting for the dog to be successful.

2.47 **a.** No, the judges wanted to show that one of Pismo's restaurants made the best chowder.
b. So that the evaluation is not swayed by personal interest.

2.49 **a.** Randomly divide volunteers into 2 groups of 50, one groups gets PH80 and the other group gets a placebo nasal spray. They are assessed before and after the treatment and any improvement in PMS symptoms measured. It would be more accurate if neither the subjects or the assessors and recorders of the PMS symptoms knew which treatment group each subject had been assigned.
b. A placebo treatment is needed to see if improvement is due to the PH80 or just a the act of spraying a liquid (with no medicinal qualities) up your nose that improves the symptoms of PMS.
c. As irritability is so subjective, double-blinding, as described in **a**, would be advisable.

Exercises 2.51 – 2.55 Answers will vary
Exercises 2.57 – 2.69

2.57 **a.** It is an observational study; no treatment was imposed.
b. If a child has attention disorder at a young age, parents who may find life difficult, may be more likely to let the child watch TV so they could have a break. The watching of TV wouldn't be causing the disorder, they would be watching TV because of the disorder!

2.59 If we are considering all people 12 years of age or older, there are probably more single people that are young, and more widowed people that are old. It tends to be the young who are at higher risk of being victims of violent crimes (for instance, staying out late at night). Hence, age could be a potential confounding variable.

2.61 Any survey conducted by mail is subject to selection bias; it eliminates any-one who doesn't have a permanent address, any-one who is on vacation or for any other reason doesn't receive mail. Once they receive the survey, many people consider surveys as junk mail and only respond to those that elicit strong feelings - resulting in non-response bias (only a few people reply).

2.63 Let us evaluate this design by considering each of the basic concepts of designing an experiment.
Replication: Each of the 8 actors was watched on tape by many of the primary care doctors.
Direct Control: The actors wore identical gowns, used identical gestures, were taped from the same position and used identical scripts.
Blocking: not used
Randomization: The article does not indicate if the 720 doctors were randomly divided into 8 groups of 90 doctors and each group randomly assigned to watch one of the actors on tape, but it is reasonable to assume this was done.
This design appears to be good because it employs many of the key concepts in designing an experiment. One possible improvement would be to randomly select the 720 primary care doctors from the entire population of primary care doctors. By randomly selecting a sample from the entire population, we can generalize our results of the study to the whole population. In this study, the conclusions only apply to this group of 720 doctors.

2.65 **a.** There are several extraneous variables, which could affect the results of the study. Two of these are subject variability and trainer variability. The researcher attempted to hold these variables constant by choosing men of about the same age, weight, body mass and physical strength and by using the same trainer for both groups. The researcher also included replication in the study. Ten men received the creatine supplement and 9 received the fake treatment. Although the article does not say, we hope that the subjects were randomly divided between the 2 treatments.

b. It is possible that the men might train differently if they knew whether they were receiving creatine or the placebo. The men who received creatine might have a tendency to work harder at increasing fat-free mass. So it was necessary to conduct the study as a blinded study.

c. If the investigator only measured the gain in fat-free mass and was not involved in the experiment in any other way, then it would not be necessary to make this a double blind experiment. However, if the investigator had contact with the subjects or the trainer, then it would be a good idea for this to be a double blind experiment. It would be particularly important that the trainer was unaware of the treatments assigned to the subjects.

2.67 **a.** There are 2 treatments in this experiment – standing or squatting and the response variable is 'amount of tip'.

b. There are a number of extraneous factors in this experiment. They include table location and how busy the place is. Blocking could be used to control for these factors. For instance, one could separately evaluate the differences in tip received between squatting and standing during busy hours and during slow hours.

c. Blocking would be essential here for the study to be successful. There are several factors that could be used for blocking. Time of day, smoking vs non-smoking sections and table location are a few possibilities. Of course, some level of blocking is already used in this study with individual waiters/waitresses as blocks.

d. All uncontrolled or unrecorded factors will be confounding variables. An additional confounding variable is economic status of individual customers.

e. The waiter flips a coin to determine whether he would stand or squat at the table. It is necessary to randomize the treatments as a strategy for dealing with extraneous variables not taken into account through direct control or blocking. We count on randomization to create 'equivalent' restaurant customers.

2.69 The response variable for the tile is whether it cracked or not in the firing process. Since two different firings will not have exactly the same temperature, tiles made from each type of clay should be fired together. Fifty tiles of each type could be used per firing. Since temperature varies within the oven, the oven should be divided into sections (blocks) where the temperature is the same within a section, but perhaps differs between sections. Then, files made from each clay type should be placed within the sections. The positions of the tiles within a section should be determined in a random fashion.

Chapter 3

Exercises 3.1 to 3.13

3.1

Choice of College Attended

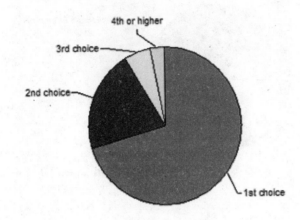

3.3

Mobile Americans

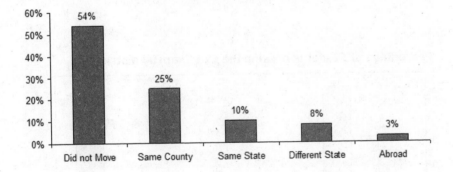

It is clear in both graphs that most Americans did not move between the years 1995 and 2000, and if they did, they stayed in the same County. It is also easy to see that the smallest proportion moved abroad. However, without the labels, it is much easier to see on the bar chart that more households moved within State than to a different State than by using a pie chart. Comparisons of categories with similar relative frequencies can be difficult to see with a pie chart.

3.5

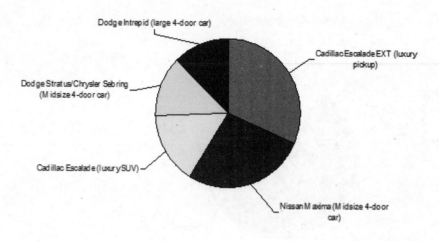

Theft Claims per 1000 vehicles

Dodge Intrepid (large 4-door car)

Dodge Stratus/Chrysler Sebring (Midsize 4-door car)

Cadillac Escalade (luxury SUV)

Cadillac Escalade EXT (luxury pickup)

Nissan Maxima (Midsize 4-door car)

3.7   a.

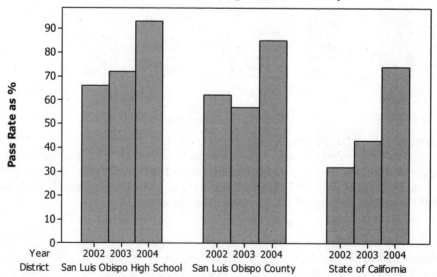

Percentage of Students passing the Exit Exam by district

Pass Rate as %

| Year | 2002 2003 2004 | 2002 2003 2004 | 2002 2003 2004 |
| District | San Luis Obispo High School | San Luis Obispo County | State of California |

b. The pass rate increased each year for San Luis Obispo High School and the State of CA from 2002 to 2004 with a sharp rise in 2004. However, in San Luis Obispo County, there was a drop in the pass rate in 2003, followed by a sharp increase in 2004 when a pass in the exam was need for graduation.

3.9   **a.**

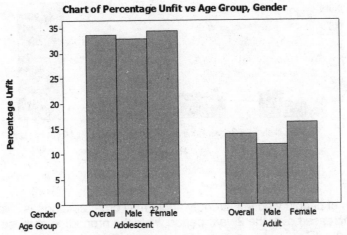

b. For both age groups, females are more unfit than the overall age group, and men are less unfit. However, this difference is much less marked in adolescents who on the whole are much more unfit than their older counterparts.

3.11   **a.**

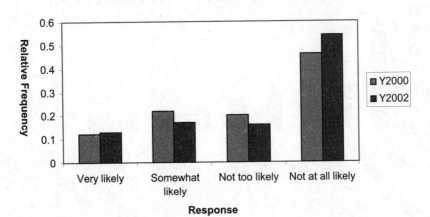

The proportion of Christmas shoppers who are "very likely" or "somewhat likely" to use the Internet has increased from 2000 to 2002, and the proportion who are "not too likely" or "not at all likely" has decreased. However it should be noted that the vast majority of Christmas shoppers (71%) are hesitant to do their Christmas shopping on-line.

b.

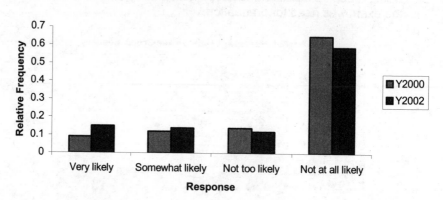

Christmas online shopping

The proportion of people who felt strongly about mail order catalogues ("very likely" or "not at all likely") increased over the 2 year period while the proportion of those who weren't too sure decreased.

3.13   **a.** There are too many categories for this pie chart to be effective.
        b.

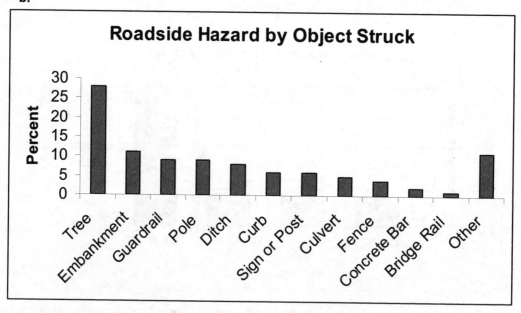

Roadside Hazard by Object Struck

This is much more effective than the pie chart. It is very easy to compare the small differences between the objects as hazards on the roads.

Exercises 3.15 to 3.21

3.15   **a.** Number of people (12 and older) who have smoked in the last month, by State

1	1 3 3 3 5 7 9
2	1 2 5 6 8 stem = hundreds (of thousands)
3	0 0 6 7 leaves = tens (of thousands)
4	5
5	4 7
6	5 8 8
7	3 4
8	6
9	0 8 9
10	2 4
11	2 7 7 8
12	1 2
13	4
14	3 7 9
15	
16	6
17	8
18	
19	
20	1
21	
22	
23	4
24	
25	
26	
27	5
28	6 7
29	
30	
31	
32	
33	6
34	
35	
36	
37	
38	
39	
40	5
41	
42	
43	
44	3
45	
46	
47	

48	
49	
50	
51	
52	
53	
54	
55	1

b. The distribution is skewed to the right with most of the states having values at the lower end of the scale. 40 out of the 50 states have less than 1,500,000 people who have smoked in the last month. There are some outliers at the high end of the distribution.

c. No, it does not indicate that tobacco is necessarily a problem in these states, NY, CA and TX are the three most heavily populated states in the US and even if they have the same proportion of smokers as others states, they will have a higher *number* of smokers because of the greater population.

d. No, it would be better to use the proportion of the population of each state that smoked during the last month. That way, the population of the state would not effect the result.

3.17

	Calorie Content (cal/100ml)of 26 Brands of Light Beers
1L	stem: tens
1H	9 leaf: ones
2L	2 3
2H	7 8 8 9 9 9
3L	0 0 1 1 1 2 2 3 3 4 5
3H	9
4L	0 1 2 3
4H	

3.19 **a.**

Very Large Urban areas		Large Urban Areas
	2	3 6 9
8	3	0 0 3 3 5 8 9
9 9	4	0 3 6 6 stem = tens
1 1 7 8	5	0 1 2 3 5 5 leaves = units
0 3 7 9	6	
2	7	
	8	
3	9	

b. Not necessarily. Philadelphia is a larger urban area than Riverside, CA, but has less extra travel time. However, *overall,* taking into account all the urban areas mentioned, or if we were to calculate the average or typical value for each type of area, then we would find that *on the whole,* the larger the urban area, the greater the extra travel time.

3.21

	High School drop out rates 1997-1999 by State
0f	5 5 5
0s	6 6 6 6 6 7 7 7 7 7 7 7 stem: tens
0*	8 8 8 8 8 8 8 9 9 9 9 9 9 9 9 9 leaf: ones
1.	0 0 0 1 1 1 1 1
1t	2 2 2 2 3 3 3 3
1f	
1s	7 7

Exercises 3.23 to 3.33

3.23   a.

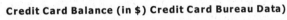

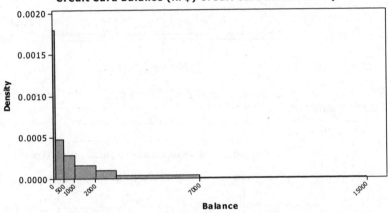

b.

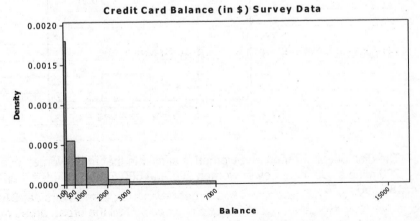

c. The results for both the survey and the credit bureau show that most college student debt is less than $2000. However, the results don't agree exactly; the credit card bureau had more students with very large debt than the survey data suggested. Maybe some of the students with very large debts were reluctant to admit to the scale of their problems!

d. Only 132 out of 1260 students replied, an 89.5% non-response rate. It may be that students with a large debt would be reluctant to respond, so this sample was not representative of all students.

3.25   **a.**

Average # Cell phone mins used per Month by Men

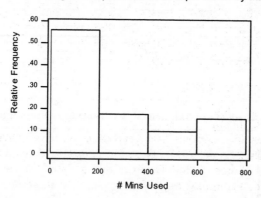

Most men average between 0 and 200 mins a month. Far fewer average between 400 and 800 mins a month.

  b.

Average # Cell phone mins used per Month by Women

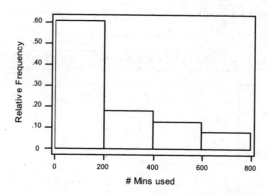

The distribution for men and women is similar in that most women average between 0 and 200 mins as do men. Fewer women average 600 to 800 minutes than men.

  c. 18 + .56 = .74

d. $.74 + \dfrac{.1}{2} = .79$

e. $\dfrac{.13 * 3}{4} + .08 = .1775$

3.27   a.

0	360, 340, 960, 530, 540, 960, 450, 500, 100, 510, 240, 396
1	280, 240, 050, 000, 320, 250, 120, 850, 670, 890, 419
2	100, 400, 120, 250, 320, 400, 460, 700, 730, 109
3	060, 330, 380, 350, 870, 150, 150
4	390, 770
5	320, 700, 220, 850, 770

stem: thousands digit
leaf: hundreds, tens and ones

The stem and leaf display suggests that a typical or representative value is in the stem 2 row, perhaps someplace in the 2230 range. The display declines as we move to higher stems and then rises at stem 5 row. There are no gaps in the display. The shape of the display is not perfectly symmetric but rather appears to stretch out quite a bit in the direction of low stems.

b.

Class Interval	Frequency	Relative Frequency
0 - <1000	12	0.2553
1000 - < 2000	11	0.2340
2000 - < 3000	10	0.2128
3000 - < 4000	7	0.1489
4000 - < 5000	2	0.0426
5000 - < 6000	5	0.1064
	47	1.0000

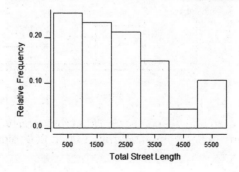

The upper tail of the histogram stretches farther than the lower tail so the histogram is positively skewed.

c. The proportion of subdivisions with total length less than 2000 is $\dfrac{12+11}{47} = \dfrac{23}{47} = .4894$ or approximately 49%. The proportion of subdivisions between 2000 and 4000 is $\dfrac{10+7}{47} = .3617$ or approximately 36%.

3.29 **a.** If the exam is quite easy, then there would be a large number of high scores with a small number of low scores. The resulting histogram would be negatively skewed.

b. If the exam were quite hard, then there would be a large number of low scores with a small number of high scores. The resulting histogram would be positively skewed.

c. The students with the better math skills would score high, while those with poor math skills would score low. This would result in basically two groups and thus the resulting histogram would be bimodal.

3.31 **a.**

Class Intervals	Frequency	Rel. Freq.	Density
.15 < .25	8	.02192	0.2192
.25 < .35	14	.03836	0.3836
.35 < .45	28	.07671	0.7671
.45 < .50	24	.06575	1.3150
.50 < .55	39	.10685	2.1370
.55 < .60	51	.13973	2.7946
.60 < .65	106	.29041	5.8082
.65 < .70	84	.23014	4.6028
.70 < .75	11	.03014	0.6028
	n = 365	1.00001	

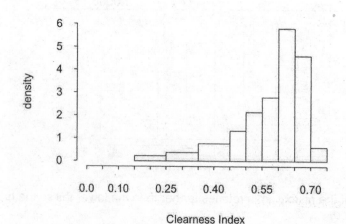

Clearness Index

b. The proportion of days with a clearness index smaller than .35 is $\dfrac{(8+14)}{365} = \dfrac{22}{365} = .06$, which converts to 6%.

c. The proportion of days with a clearness index of at least .65 is $\dfrac{(84+11)}{365} = \dfrac{95}{365} = .26$, which converts to 26%.

3.33

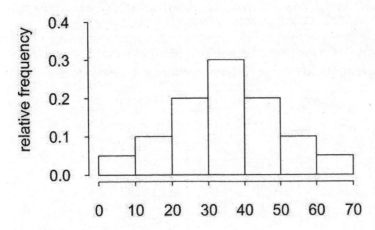

This histogram is symmetric.

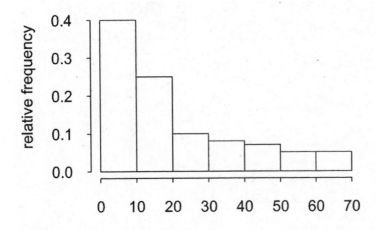

This histogram is positively skewed.

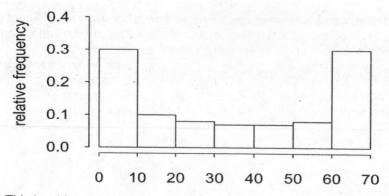

This is a bimodal histogram. While it is not perfectly symmetric it is close to being symmetric.

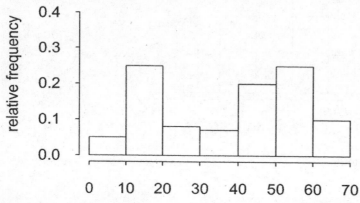

This is a bimodal histogram.

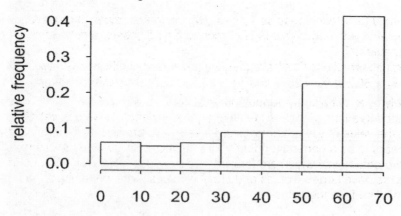

This is a negatively skewed histogram.

Exercises 3.35 to 3.45

3.35   **a.** There are several values that have identical or nearly identical x-values yet different y-values. Therefore, the value of y is not determined solely by x , but also by various other factors. There appears to be a tendency for y to decrease as x increases.

        b. People with low body weight tend to be small people and it is possible their livers may be smaller than the liver of an average person. Conversely, people with high weight tend to be large people and their livers may be larger than the liver of an average person. Therefore, we would expect the graft weight ratio to be large for low weight people and small for high weight people.

3.37   **a.**

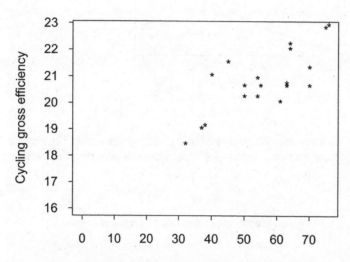

Percentage of type I (slow twitch) muscle fibers

        There is a tendency for y to increase as x does. That is, larger values of gross efficiency tend to be associated with larger values of percentage type I fibers (a positive relationship between the variables).

        b. There are several observations that have identical x-values yet different y-values (for example, $x_6 = x_7 = 50, but\ y_6 = 20.5\ and\ y_7 = 20.1$). Thus, the value of y is *not* determined solely by x, but also by various other factors.

3.39   There are several observations that have identical or nearly identical x-values yet different y-values. Therefore, the value of y is not determined solely by x, but also by various other factors. There appears to be a general tendency for y to decrease in value as x increases in value. There are two data points which are far removed from the remaining data points. These two data points have large x-values and small y-values. Their presence might have an undue influence on a line fit to the data.

3.41 **a.**

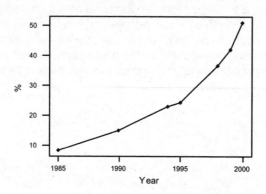

% of Households with a Computer

b. There has been an increase in the ownership of computers over time since 1985. At first the increase was slow and then from 1995 the increase has been increasing at a more rapid rate.

3.43 **a.**

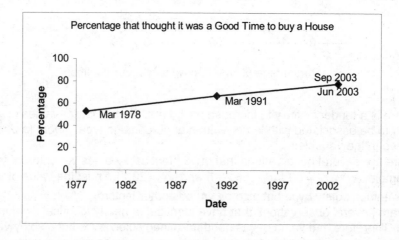

b.

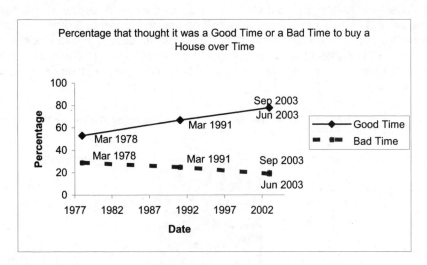

c. The time series plot give a better trend over time as it shows a time scale, which the bar chart does not.

3.45 In both 2001 and 2002 the box office sales dropped in Weeks 2, 6 and in the last two weeks of the summer. The seasonal peaks occurred during Weeks 4, 9 and 13.

Exercises 3.47 to 3.63

3.47

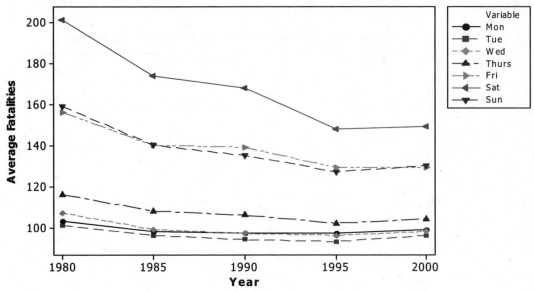

For those students who have English as a first Language there is very little difference between the average Verbal and Math scores. For students who speak English and other language, they do less well in both subjects, but it is more noticeable in the Verbal scores. Those students who speak a language other than English as their first language score on average about 55 points less than those who speak only English. However, their average Math scores are as high as the English only speaking students.

3.49

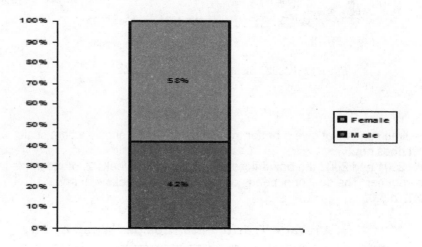

Where are the Men at Community College?

3.51   a.

Av. Transp. Exp. for a British household (in Pounds Sterling)

b.

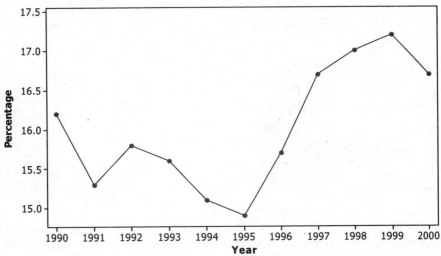

Percentage of Household expenses spent on Transportation

The time series plots form part a. do agree with the statement. It is clear from the first graph that the actual expenditure has been increasing. Although the percentage of household expenditure looks volatile, in the 10 years of this study, it has varied from 14.9% to 17.2%, small compared with the increase in actual expenditure.

3.53   a.

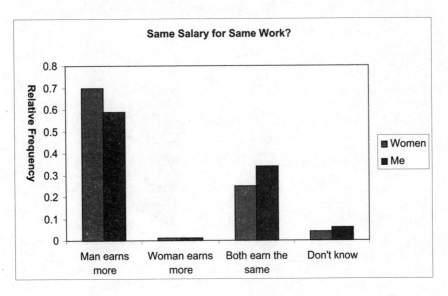

Same Salary for Same Work?

b.

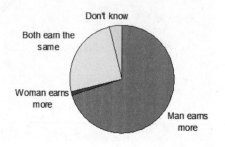

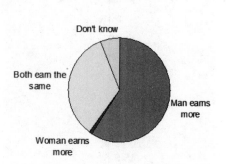

c. It's much easier to compare the differences using the comparative bar chart. The categories are next to each other and any difference is easy to see. To compare the size of the slice between two pie charts is more difficult.

d. The majority of both men and women think that men earn more money for the same work. Very few men or women think that a woman earns more. More women than women think that a man earns more or earns the same as woman.

3.55   a.

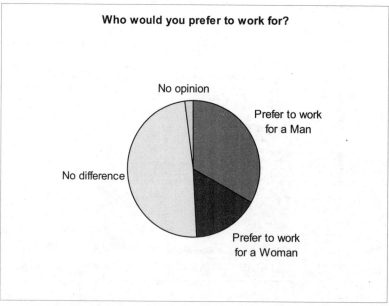

Given the choice, most people don't mind whether they work for a man or a woman. Of those that do have an opinion, most would prefer to have a male boss.

b.

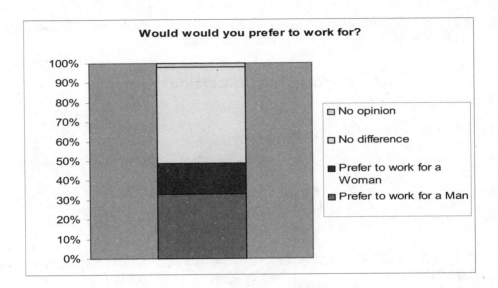

3.57

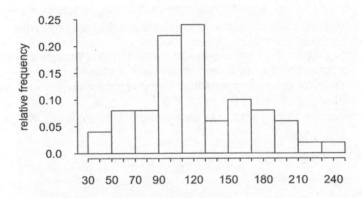

3.59

0	6 7 5 8 9 0 0 5 0
1	8 6 1 5 2 1 6 7 0 6 3 1 2 2 8 4 3 1 8 0 8 7 8 7 7
2	5 6 9 0 7 8 6 9 0 6 1 9 7 1
3	7 5 5 0 0 0 1 5 0 5
4	5 6 7 8 2 3 0
5	
6	7 1 stem: tens digit
7	0 leaf: ones digit

The data values are concentrated between 0 to 40, with a few larger values. Overall, the plot appears to be skewed to the right.

3.61 If 39% of those with critical housing needs lived in urban areas and 42% lived in suburban areas, then 19% (100 − 39 − 42) lived in rural areas.

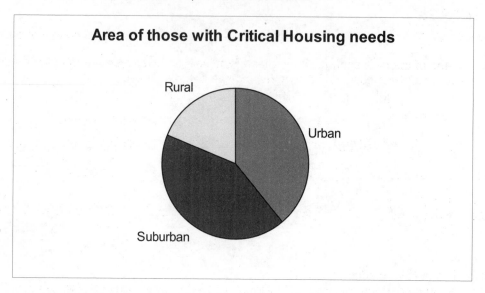

Area of those with Critical Housing needs

3.63 **a.** The two histograms do give different impressions about the distribution of values. For the first histogram, it appears that more frozen meals have sodium content around 800mg. However, the second histogram suggests that sodium content is fairly uniform from 300mg to 900mg and then drops off above 900mg.

b. Using the first histogram, the proportion of observations that are less than 800 is approximately

$$\frac{6+7+\dfrac{10}{2}}{(6+7+10+4)} = \frac{18}{27} = .6667.$$

Using the second histogram, the proportion of observations that are less than 800 is approximately

$$\frac{6+5+5+\left(\dfrac{800-750}{150}\right)(6)}{(6+5+5+6+4+1)} = \frac{16+2}{27} = \frac{18}{27} = .6667.$$

The actual proportion is $\dfrac{18}{27} = \dfrac{2}{3} = .6667.$

Chapter 4

Exercises 4.1 to 4.13

4.1   mean = $\dfrac{9707}{10} = 970.7$ median = $\dfrac{707 + 769}{2} = \dfrac{1476}{2} = 738$. There are a couple of outliers at
the high end of the data set and so the mean is influenced by these more than the median.
For this reason, the median is more representative of the sample.

4.3   **a.**

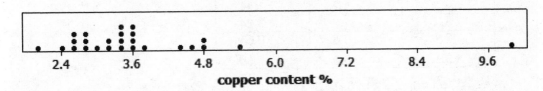

% of copper content for a sample of Bidri artifacts

b. Mean = 3.645%, median = 3.35%.
c. An 8% trimmed mean will drop the outlier of 10.1% from the high end of the values. This
   will make the 8% trimmed mean smaller than the mean.

4.5   **a.** Mean = 448.3 fatalities.
b. Median = 446 fatalities.
c. The 20 days that had the highest number of fatalities are not a random sample of the 365
   days of the year and so cannot be used to generalize to the rest of the year.

4.7   **a.** No, we have no knowledge of the costs of any intermediate plans.
b. The number of plans is not given in the data set.
c. Iowa, Minnesota, Montana, Nebraska, North and South Dakota and Wyoming.
d. Florida.
e. Alaska
f. $9.0925

4.9   Expensive house do not pull the median price up. Sales of very expensive houses will
increase the mean price, just as the sales of very inexpensive houses will decrease the mean
price. If there are more sales of either very high or very low priced homes, this will affect the
median price. More higher priced sales than lower priced sales, the median increases; more
lower priced sales, the median price decreases.

4.11   The four observations ordered are: 27.3, 30.5, 33.5, 36.7.
The median = 32 and the mean = 128/4 = 32. If we add the value 32 to the data set, the new
ordered data set ordered is 27.3, 30.5, 32, 33.5, 36.7. For this data set, the sample median is
32 and the mean is $\bar{x} = \dfrac{160}{5} = 32$.

4.13 The median and trimmed mean (trimming percentage of at least 20) can be calculated.

$$\text{sample median} = \frac{57 + 79}{2} = \frac{136}{2} = 68$$

$$20\% \text{ trimmed mean} = \frac{35 + 48 + 57 + 79 + 86 + 92}{6} = \frac{397}{6} = 66.17$$

Exercises 4.15 to 4.27

4.15 $n = 5$, $\Sigma x = 6$, $\overline{x} = \frac{6}{5} = 1.2$.

x	$(x - \overline{x})$	$(x - \overline{x})^2$
1	0.2	.04
0	1.2	1.44
0	1.2	1.44
3	1.8	3.24
2	0.8	0.64
6	0.0	6.80

$$s^2 = \frac{6.8}{4} = 1.70, \quad s = \sqrt{1.70} = 1.3038$$

4.17 It would have a large standard deviation. Some parents would spend much more than the average and some parents would either not be able to afford to spend that much (or not be willing to) or would rather spread it over the school year.

4.19 **a.** standard deviation = 50.058 fatalities.
b. Smaller. There is less variability in the data for Memorial Day.
c. There is less variability for holidays that always occur on the same days of the week. Memorial Day, Labor Day and Thanksgiving Day have standard deviations of 18.2, 17.7 and 15.3 respectively whereas New Year's Day, July 4th and Christmas Day have standard deviations of 50.1, 47.1 and 52.4.

4.21 **a.** The sample size for Los Osos and Morro Bay are different so the average for the combined areas is $\frac{(606456)(114) + (511866)(123)}{(114 + 123)}$.
b. Because Paso Robles has the bigger range of values, it is likely to have the larger standard deviation.
c. Because the difference in the High prices is so much greater than the difference in the Low prices, it would suggest that the distribution of house prices in Paso Robles is right skewed and the mean price is higher than the median suggesting the median would be lower in Paso Robles than in Grover Beach.

4.23 mean = $\frac{20179}{27} = 747.37$ standard deviation = 606.894 (both in thousands of $)

mean + 2 standard deviations = 747.37 + 2(606.894) = 1961.158 (or $1,961,158)

4.25 26 data points.
1st quartile will be the middle of the first 13 ordered data points; i.e. the 7th point: 10478
3rd quartile will be the middle of the second 13 ordered data points; i.e. the 20th point: 11778
Interquartile range: 1300

4.27   **a.** 31 data points.

1st quartile will be the middle of the first 15 ordered data points; i.e. the 8th point: 51

3rd quartile will be the middle of the second 15 ordered data points; i.e. the 24th point: 69

Interquartile range: 18

b. The IQR for inpatient cost-to-charge ratio in Example 4.9 is 14. There is more variability in cost-to-charge ratios for outpatient services.

Exercises 4.29 to 4.35

4.29   **a.** As the mean (22.4) is greater than the median (18), the distribution is positively skewed.

     b.

Travel Time to Work

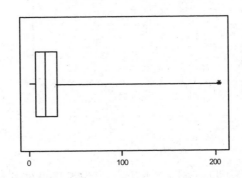

c. The maximum point (205) is greater than the upper quartile + 1.5 times the interquartile range (31 + 1.5(24)). Therefore there is at least one extreme one outlier, possibly more and possibly some mild outliers.

4.31   **a.** mean = 59.85 hours, median = 55.0 hours. As the mean is greater than the median, the distribution is likely to be skewed to the right.

     b.

Extra Travel Time for Commuters

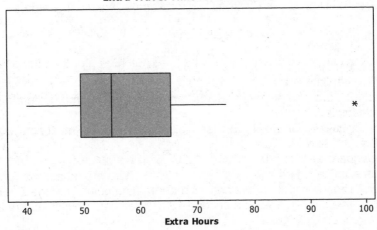

Extra Hours

35

The number of extra commuter hours for urban areas is centered around 55 hours and the distribution is skewed to the right. There is a major outlier, Los Angeles at 98 hours. Excluding the outlier the range is from 40 hours to 75 hours.

4.33 **a.** African-American: The mean blood lead level (4.93) is higher than the median (3.6). This is due to the two unusually large values in the data set that have an effect on the mean but not on the median.

b.

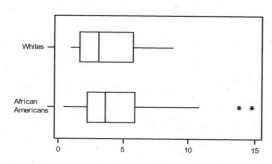

Blood Lead Level

Overall, African Americans have a higher blood lead level than Whites and also had greater variability. There were 2 samples from the African Americans that had much higher blood lead levels than the others of the same ethnicity.

4.35 **a.** For the excited delirium sample, the median is 0.4. The lower quartile is 0.1 and the upper quartile is 2.8. The interquartile range is 2.7.

For the No Excited Delirium sample, the median is $\dfrac{1.5+1.7}{2}$ = 1.6. The lower quartile is 0.3 and the upper quartile is 7.9. The interquartile range is 7.6.

b. To check for outliers for the Excited Delirium sample, we must compute
1.5 x 2.7 = 4.05 and 3 x 2.7 = 8.1
The lower quartile – 4.05 = 0.1 – 4.05 = -3.05
The upper quartile + 4.05 = 2.8 + 4.05 = 6.85
The lower quartile – 8.1 = 0.1 – 8.1 = -8.0
The upper quartile + 8.1 = 2.8 + 8.1 = 10.9
So 8.9 and 9.2 are mild outliers and 11.7 and 21.0 are extreme outliers for the Excited Delirium sample.
To check for outliers for the No Excited Delirium sample, we must compute
1.5 x 7.6 = 11.4 and 3 x 7.6 = 22.8
The lower quartile –11.4 = 0.3-11.4 = -11.1
The upper quartile + 11.4 = 7.9 + 11.4 = 19.3   There are no outliers for this sample.

c.

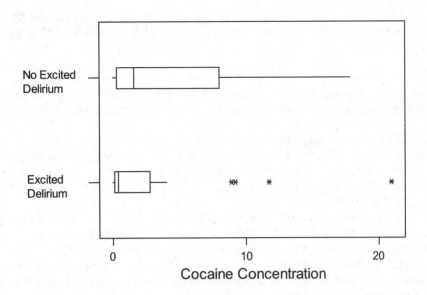

Cocaine Concentration

The median marker for each boxplot is located at the lower end of the boxplot. The boxplot for the No Excited Delirium is much wider than for Excited Delirium indicating much more variability in the middle half of the data. Both boxplots have short whiskers on the lower end and long whiskers on the higher end. The Excited Delirium sample has 2 mild outliers and 2 extreme outliers. There are no outliers for the No Excited Delirium sample.

Exercises 4.37 to 4.49

4.37   **a.** The value 57 is one standard deviation above the mean. The value 27 is one standard deviation below the mean. By the empirical rule, roughly 68% of the vehicle speeds were between 27 and 57.

b. From part **a** it is determined that 100%   68% = 32% were either less than 27 or greater than 57. Because the normal curve is symmetric, this allows us to conclude that half of the 32% (which is 16%) falls above 57. Therefore, an estimate of the percentage of fatal automobile accidents that occurred at speeds over 57 mph is 16%.

4.39   **a.** The value 59.60 is two standard deviations above the mean and the value 14.24 is two standard deviations below the mean. By Chebyshev's rule the percentage of observations between 14.24 and 59.60 is at least 75%.

b. The required interval extends from 3 standard deviations below the mean to 3 standard deviations above the mean. From 36.92   3(11.34) = 2.90 to 36.92 + 3(11.34) = 70.94.

c. $\bar{x} - 2s$ = 24.76   2(17.20) =   9.64
In order for the histogram for NO_2 concentration to resemble a normal curve, a rather large percentage of the readings would have to be less than 0 (since $\bar{x}$ 25 =   9.64). Clearly, a reading cannot be negative, hence the histogram cannot have a shape that resembles a normal curve.

4.41 For the first test the student's z-score is $\frac{(625-475)}{100} = 1.5$ and for the second test it is

$\frac{(45-30)}{8} = 1.875$. Since the student's z-score is larger for the second test than for the first

test, the student's performance was better on the second exam.

4.43 Since the histogram is well approximated by a normal curve, the empirical rule will be used to obtain answers for parts **a** – **c**.

a. Because 2500 is 1 standard deviation below the mean and 3500 is 1 standard deviation above the mean, about 68% of the sample observations are between 2500 and 3500.

b. Since both 2000 and 4000 are 2 standard deviations from the mean, approximately 95% of the observations are between 2000 and 4000. Therefore about 5% are outside the interval from 2000 to 4000.

c. Since 95% of the observations are between 2000 and 4000 and about 68% are between 2500 and 3500, there is about 95 – 68 = 27% between 2000 and 2500 or 3500 and 4000. Half of those, 27/2 = 13.5%, would be in the region from 2000 to 2500.

d. When applied to a normal curve, Chebyshev's rule is quite conservative. That is, the percentages in various regions of the normal curve are quite a bit larger than the values given by Chebyshev's rule.

4.45 The recorded weight will be within 1/4 ounces of the true weight if the recorded weight is

between 49.75 and 50.25 ounces. Now, $\frac{(50.25-49.5)}{.1} = 7.5$ and $\frac{(49.75-49.5)}{.1} = 2.5$. Also,

at least $1 - 1/(2.5)^2 = 84\%$ of the time the recorded weight will be between 49.25 and 49.75. This means that the recorded weight will exceed 49.75 no more than 16% of the time. This implies that the recorded weight will be between 49.75 and 50.25 no more than 16% of the time. That is, the proportion of the time that the scale showed a weight that was within 1/4 ounce of the true weight of 50 ounces is no more than 0.16.

4.47 Because the number of answers changed from right to wrong cannot be negative and because the mean is 1.4 and the value of the standard deviation is 1.5, which is larger than the mean, this implies that the distribution is skewed positively and is not a normal curve. Since $(6 - 1.4)/1.5 = 3.07$, by Chebyshev's rule, at most $1/(3.07)^2 = 10.6\%$ of those taking the test changed at least 6 from correct to incorrect.

4.49 **a.**

Class	Frequency	Rel Freq.	Cum. Rel. Freq.
5 < 10	13	.26	.26
10 < 15	19	.38	.64
15 < 20	12	.24	.88
20 < 25	5	.10	.98
25 < 30	1	.02	1.00
	n = 50	1.00	

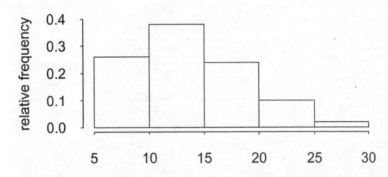

1989 per capita expenditures on libraries

b. i. The 50th percentile is between 10 and 15 since the cumulative relative frequency at 10 is .26 and at 15 it is .64. The 50th percentile is approximately
10 + 5(50 − 26)/38 = 10 + 3.158 = 13.158.

ii. The 70th percentile is between 15 and 20 since the cumulative relative frequency at 15 is .64 and at 20 it is .88. The 70th percentile is approximately
15 + 5(70 − 64)/24 = 15 + 1.25 = 16.25.

iii. The 10th percentile is between 5 and 10 and is approximately
5 + 5(10 − 0)/26 = 5 + 1.923 = 6.923.

iv. The 90th percentile is between 20 and 25 and is approximately
20 + 5(90 − 88)/10 = 20 + 1 = 21.

v. The 40th percentile is between 10 and 15 and is approximately
10 + 5(40 − 26)/38 = 10 + 1.842 = 11.842.

Exercises 4.51 to 4.69

4.51 The sample median determines this salary. Its value is $\dfrac{4443 + 4129}{2} = \dfrac{8572}{2} = 4286.$

The mean salary paid in the six counties is
$$\frac{5354 + 5166 + 4443 + 4129 + 2500 + 2220}{6} = \frac{23812}{6} = 3968.67.$$
Since the mean salary is less than the median salary, the mean salary is less favorable to the supervisors.

39

4.53   **a.** The mean is $1,110,766 which is larger than the median which is $275,000. There are probably a few baseball players who earn a lot of money (in the millions of dollars) while the bulk of baseball players earn in the hundreds of thousands of dollars. The outliers have a greater influence on the mean value making the value larger than the median.

b. If the population was all 1995 salaries and all the salaries were used to compute the mean then the reported mean was the population mean .

4.55   **a.** $\bar{x} = \dfrac{532}{11} = 48.364$

b.

Observation	Deviation	(Deviation)^2
62	13.636	185.9405
23	25.364	643.3325
27	21.364	456.4205
56	7.636	58.3085
52	3.636	13.2205
34	14.364	206.3245
42	6.364	40.5005
40	8.364	69.9565
68	19.636	385.5725
45	3.364	11.3165
83	34.636	1199.6525
	0.004	3270.5455

$s^2 = (3270.5455)/10 = 327.05$ and $s = \sqrt{327.05} = 18.08$. s^2 could be interpreted as the mean squared deviation from the average distance when detection takes place. This is 327.05 squared centimeters. The standard deviation s could be interpreted as the typical amount by which a distance deviates from the average distance when detection first takes place. This is 18.08 centimeters.

4.57   Multiplying each data point by 10 yields:

x	$(x - \bar{x})$	$(x - \bar{x})^2$
620	136.36364	18595.04132
230	253.63636	64331.40496
270	213.63636	45640.49587
560	76.36364	5831.40496
520	36.36364	1322.31405
340	143.63636	20631.40496
420	63.63636	4049.58678
400	83.63636	6995.04132
680	196.36364	38558.67769
450	33.63636	1131.40496
830	346.36364	119967.76860
	0.00004	327054.54545

$$\bar{x} = 483.63636$$

$$s^2 = \frac{(327054.54545)}{10} = 32705.4545 \qquad\qquad s = \sqrt{32705.4545} = 180.846$$

The standard deviation for the new data set is 10 times larger than the standard deviation for the original data set.

4.59 a. median: 20.88 1st Quartile: 18.09 3rd Quartile: 22.20

b. iqr = 22.2 − 18 09 = 4.11

1.5(iqr)= 1.5(4.11) = 6.165 Anything below 18.09 − 6.165 = 11.925 or above
22.2 + 6.165 = 28 365 would be an outlier. There are 2 outliers, 35. 78 and 36.73.

c.

Oxygen capacity (mL/kg/min) for Male middle-aged Runners

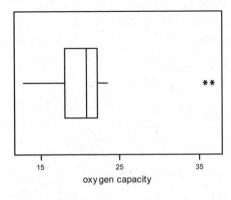

oxygen capacity

Most of the subjects have an oxygen capacity of between 20 and 25 mL/kg/min. There are 2 subjects with much higher oxygen capacities; of over 35 mL/kg/min.

4.61 a. $\bar{x} = \frac{2696}{14} = 192.57$

The data arranged in ascending order is: 160, 174, 176, 180, 180, 183, 187, 191, 194, 200, 205, 211, 211, 244.

The median is the average of the two middle values since $n = 14$, an even number. The median value is $\frac{(187+191)}{2} = 189$.

b. The median value would remain unchanged, but the mean value would decrease. The new mean value is $\frac{2656}{14} = 189.71$.

c. The trimmed mean =

$\frac{174+176+180+180+183+187+191+194+200+205+211+211}{12} = \frac{2292}{12} = 191.$ The

trimming percentage is $\frac{1}{14}$ x 100 = 7.14%

41

d. If the largest observation is 204, then the new trimmed mean =
$$\frac{174+176+180+180+183+187+191+194+200}{9} = \frac{1665}{9} = 185$$

If the largest observation is 284, the trimmed mean is the same as in **c.**

4.63   The median is $\dfrac{119+119}{2} = 119$.

The lower quartile is 87.
The upper quartile is 182.
The interquartile range is 182 – 87 = 95.
To check for outliers, we calculate
1.5 × iqr = 1.5× 95 = 142.5
3.0 × iqr = 3.0× 95 = 285
The upper quartile + 1.5 × iqr = 324.5
The lower quartile – 1.5 × iqr = -55.5
The upper quartile + 3 × iqr = 467
The lower quartile – 3× iqr = -198
Therefore, 511 is an extreme outlier.

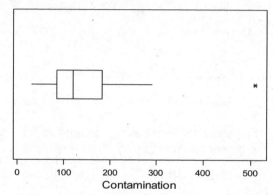

Contamination

The median line is not at the center of the box so there is a slight asymmetry in the middle half of the data. The upper whisker is slightly longer than the lower whisker. There is one extreme outlier, 511.

4.65   Budget: lower quartile = (4.0 + 4.1)/2 = 4.05
upper quartile = (7.7 + 7.9)/2 = 7.80
iqr = 7.80   4.05 = 3.75
lower mild outlier cutoff = 4.05   1.5(3.75) = 1.575
upper mild outlier cutoff = 7.80 + 1.5(3.75) = 13.425
There are no outliers in the budget data.
Midrange : lower quartile = 6.7
upper quartile = 8.1
iqr = 8.1   6.7 = 1.4
lower mild outlier cutoff = 6.7   1.5(1.4) = 4.6
upper mild outlier cutoff = 8.1 + 1.5(1.4) = 10.2
lower extreme outlier cutoff = 6.7   3(1.4) = 2.5

There are no outliers on the high end. There are two outliers on the low end. A mild outlier of 4 and an extreme outlier of 1.5.

First-class: lower quartile = 6.6

upper quartile = 7.8

iqr = 7.8   6.6 = 1.2

lower mild outlier cutoff = 6.6   1.5(1.2) = 4.8

upper mild outlier cutoff = 7.8 + 1.5(1.2) = 9.6

lower extreme outlier cutoff = 6.6   3(1.2) = 3.0

upper extreme outlier cutoff = 7.8 + 3(1.2) = 11.4

There is an extreme outlier of 1.8 on the low end and a mild outlier of 9.6 on the high end.

Boxplots for the franchise costs of the three types of hotels are given below.

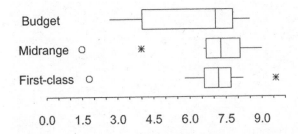

Franchise cost as a percentage of total room revenue

The median franchise cost is about the same for each type of hotel. The variability of franchise cost differs substantially for the three types. With the exception of a couple of outliers, first-class hotels vary very little in the cost of franchises, while budget hotels vary a great deal in regard to franchise costs. Ignoring the outliers, the distribution of franchise costs for first-class hotels is quite symmetric; mid-range hotels have a distribution of franchise costs which is slightly skewed to the right; while budget hotels have a distribution of franchise costs which is skewed to the left.

4.67   Since the mean is larger than the median, this suggests that the distribution of values is positively skewed or has some outliers with very large values.

4.69   **a.** The 16th and 84th percentiles are the same distance from the mean but on opposite sides. Since 80 is 20 units below 100, the 84th percentile is 20 units above 100, which would be at 120.

b. Since 84   16 = 68, roughly 68% of the scores are between 80 and 120. Thus, by the empirical rule, 120 is one standard deviation above the mean, so the standard deviation has value 20.

c. The value 90 would have a z-score of (90   100)/20 =   1/2.

d. The value 140 would be two standard deviations above the mean and thus roughly .5(5%)=2.5% of the scores would be larger than 140. Hence 140 would be the 97.5 percentile.

e. The value 40 is three standard deviations below the mean, so only about .5(.3%) or .15% of the scores are below 40. There would not be many scores below 40.

Chapter 5

Exercises 5.1 – 5.15

5.1
  a. A positive correlation would be expected, since as temperature increases cooling costs would also increase.
  b. A negative correlation would be expected, since as interest rates climb fewer people would be submitting applications for loans.
  c. A positive correlation would be expected, since husbands and wives tend to have jobs in similar or related classifications. That is, a spouse would be reluctant to take a low-paying job if the other spouse had a high-paying job.
  d. No correlation would be expected, because those people with a particular I.Q. level would have heights ranging from short to tall.
  e. A positive correlation would be expected, since people who are taller tend to have larger feet and people who are shorter tend to have smaller feet.
  f. A weak to moderate positive correlation would be expected. There are some who do well on both, some who do poorly on both, and some who do well on one but not the other. It is perhaps the case that those who score similarly on both tests outnumber those who don't.
  g. A negative correlation would be expected, since there is a fixed amount of time and as time spent on homework increases, time in watching television would decrease.
  h. No correlation overall, because for small or substantially large amounts of fertilizer yield would be small.

5.3

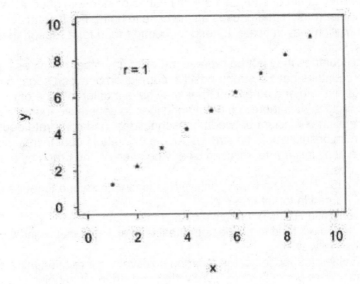

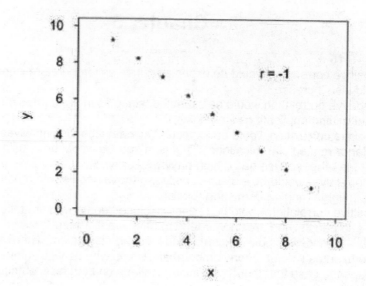

r = -1

5.5   **a.**

$$r = \frac{\sum z_x z_y}{n-1} = \frac{4.720}{5} = 0.944$$

The correlation is strong and positive.

b. Increasing sugar consumption doesn't cause or lead to higher rates of depression, it may be another reason that causes an increase in both. For instance, a high sugar consumption may indicate a need for comfort food for a reason that also causes depression.

c. These countries may not be representative of any other countries. It may be that only these countries have a strong positive correlation between sugar consumption and depression rate and other countries may have a different type of relationship between these factors. It is therefore not a good idea to generalize these results to other countries.

5.7   The correlation coefficient between household debt and corporate debt would be positive, and the relationship would be strong. As the household debt increases, the corporate debt increases at a similar rate; this can clearly be seen on the graph by a constant width between the two lines.

5.9   **a.** $r = 0.335$. There is a weak positive relationship between timber sales and the amount of acres burned in forest fires.

b. No. Correlation does not imply a causal relationship.

5.11   **a.** Since the points tend to be close to the line, it appears that x and y are strongly correlated in a positive way.

b. An r value of .9366 indicates a strong positive linear relationship between x and y.

c. If x and y were perfectly correlated with $r = 1$, then each point would lie exactly on a line. The line would not necessarily have slope 1 and intercept 0.

5.13 $r = \dfrac{27918 - \dfrac{(9620)(7436)}{2600}}{\sqrt{36168 - \dfrac{(9620)(9620)}{2600}}\sqrt{23145 - \dfrac{(7436)(7436)}{2600}}}$

$= \dfrac{27918 - 27513.2}{\sqrt{574}\sqrt{1878.04}} = \dfrac{404.8}{(23.9583)(43.336)} = .3899$

There is a weak positive linear relationship between high school GPa and first-year college GPA.

5.15 No. An r value of 0.085 indicates an extremely weak relationship between support for environmental spending and degree of belief in God.

Exercises 5.17 – 5.35

5.17 a.

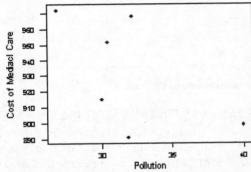

There is a weak negative association between pollution and the cost of medical care.

b. x = Pollution and y = cost; x = 191.1, x^2 = 6184.05 xy = 177807,
y = 5597, n = 6, $\bar{x} = 31.85, \bar{y} = 932.833$

$S_{xy} = \sum xy - \dfrac{(\sum x)(\sum y)}{n} = 177807 - \dfrac{(191.1)(5597)}{6} = -457.45$

$S_{xx} = \sum x^2 - \dfrac{(\sum x)^2}{n} = 6184.05 - \dfrac{(191.1)^2}{6} = 97.515$

The slope, $b = \dfrac{S_{xy}}{S_{xx}} = \dfrac{-457.45}{97.515} = -4.69$

The intercept, $a = \bar{y} - b\bar{x} = 932.833 - (-4.69)(31.85) = 1082.21$

The equation: $\hat{y}$ = 1082.21 - 4.69x

c. The slope is negative, consistent with the description in part a.

d. Yes, it does support the conclusion that elderly people that live in more polluted areas have higher medical costs, but care must be taken not to state that the pollution *causes* the high medical costs – or even the high medical costs causes the pollution!

5.19 **a.** The dependent variable is the number of fruit and vegetable servings per day. The independent variable is the number of hours of TV viewed per day.

     b. Negative, because as the number of hours of TV watched *increases*, the number of serving of fruit and vegetables *decreases*.

5.21 **a.**

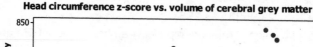

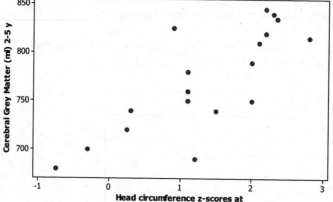

b. $n = 18, \Sigma x = 24.35, \Sigma y = 13890\ \Sigma x^2 = 49.6775, \Sigma y^2 = 10767400, \Sigma xy = 19501.75$

$\bar{x} = 1.3528, \bar{y} = 771.667$

$$S_{xy} = 19501.75 - \frac{(24.35)(13890)}{18} = 711.667$$

$$S_{xx} = 49.6775 - \frac{(24.35)^2}{18} = 16.737$$

$$S_{yy} = 10767400 - \frac{13890^2}{18} = 48950$$

$$r = \frac{S_{xy}}{\sqrt{S_{xx}}\sqrt{S_{yy}}} = \frac{711.677}{\sqrt{16.737}\sqrt{48950}} = .7863$$

c. The slope, $b = \dfrac{S_{xy}}{S_{xx}} = \dfrac{711.667}{16.737} = 42.521$

    The intercept, $a = \bar{y} - b\bar{x} = 771.667 - (42.521)(1.3528) = 714.145$

    The equation: $\hat{y} = 714.145 + 42.521\,x$

d. When head circumference z-score is 1.8, the predicted volume of grey matter is 790.68 ml.

e. The least squares line was calculated using values of z-scores of between -0.75 and 2.8 and therefore is only valid for values in this range. We don't know if the relationship between cerebral grey matter and head circumference z-score remains the same outside these values and so this equation cannot be used for prediction.

5.23   a.

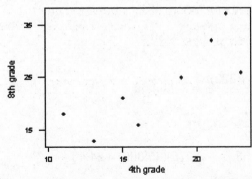

There is a moderately strong positive linear relationship between the percentage of public schools who were at or above the proficient level in math in 4th and 8th grade in the 8 states.

b. x = 4th grade and y = 8th grade; x = 140, x^2 = 2586 xy = 3497,
y = 188, n = 8, $\bar{x} = 17.5, \bar{y} = 23.5$

$$S_{xy} = \sum xy - \frac{(\sum x)(\sum y)}{n} = 3497 - \frac{(140)(188)}{8} = 207$$

$$S_{xx} = \sum x^2 - \frac{(\sum x)^2}{n} = 2586 - \frac{(140)^2}{8} = 136$$

The slope, $b = \dfrac{S_{xy}}{S_{xx}} = \dfrac{207}{136} = 1.522$

The intercept, $a = \bar{y} - b\bar{x} = 23.5 - 1.522(17.5) = -3.135$

The equation: $\hat{y}$ = -3.135 + 1.522x

c. Predicted 8th grade = -3.135 + 1.522(4th grade percent)   -3.135 + 1.522(14) = 18 (rounded to nearest integer). This is 2% lower than the actual 8th grade value of 20 for Nevada.

5.25   a.

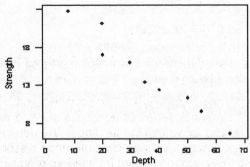

There appears to be a negative linear association between carbonation depth and the strength of concrete for a sample of core specimens.

b. $x = 323,$ $x^2 = 14339,$ $xy = 3939.9,$
$y = 130.8,$ $n = 9,$ $\bar{x} = 35.889,$ $\bar{y} = 14.533$

$$S_{xy} = \sum xy - \frac{\left(\sum x\right)\left(\sum y\right)}{n} = 3939.9 - \frac{(323)(130.8)}{9} = -754.367$$

$$S_{xx} = \sum x^2 - \frac{\left(\sum x\right)^2}{n} = 14339 - \frac{(323)^2}{9} = 2746.889$$

The slope, $b = \dfrac{S_{xy}}{S_{xx}} = \dfrac{-754.367}{2746.889} = -0.275$

The intercept, $a = \bar{y} - b\bar{x} = 14.533 - (-0.275)(35.889) = 24.40$

The equation: $\hat{y} = 24.4 - 0.275x$

c. When depth is 25, predicted strength $= 24.4 - 0.275(25) = 17.5$

d. The least squares line was calculated using values of "depth" of between 8 mm and 65 mm and therefore is only valid for values in this range. We don't know if the relationship between depth and strength remains the same outside these values and so this equation cannot be used.   A depth of 100 mm is clearly outside these values and it would be unreasonable to use this equation to predict strength.

5.27 The slope is the average increase in the y variable for an increase of one unit in the x variable. Because the home prices (y variable) dropped by an average of $4000 (-4000) for every (1) mile (x variable) from the Bay area, the slope is -4000/1 = -4000.

5.29 **a.** $x = 240,$ $x^2 = 6750$ $xy = 199750,$
$y = 7250,$ $n = 11,$ $\bar{x} = 21.818, \bar{y} = 659.091$

$$S_{xy} = \sum xy - \frac{\left(\sum x\right)\left(\sum y\right)}{n} = 199750 - \frac{(240)(7250)}{11} = 41568.182$$

$$S_{xx} = \sum x^2 - \frac{\left(\sum x\right)^2}{n} = 6750 - \frac{(240)^2}{11} = 1513.636$$

The slope, $b = \dfrac{S_{xy}}{S_{xx}} = \dfrac{41568.182}{1513.636} = 27.462$

The intercept, $a = \bar{y} - b\bar{x} = 659.091 - 27.462(21.818) = 59.925$

The equation: $\hat{y} = 59.925 + 27.462x$

b. Concentration with 18% bare ground: $59.925 + 27.462(18) = 554$ (to nearest integer)

c. No, because the data used to obtain the least squares equation was from steeply sloped plots, so it would not make sense to use it to predict runoff sediment from gradually sloped plots. You would need to use data from gradually sloped plots to create a least squares regression equation to predict runoff sediment from gradually sloped plots.

5.31 **a.**

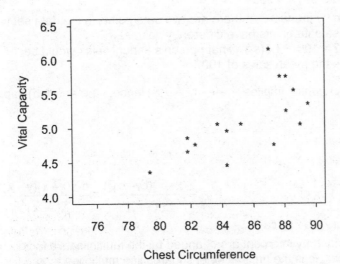

The graph reveals a moderate linear relationship between x and y.

b. $b = \dfrac{6933.48 - \dfrac{(1368.1)(80.9)}{16}}{117123.85 - \dfrac{(1368.1)^2}{16}} = \dfrac{6933.48 - 6917.456}{117123.85 - 116981.101} = \dfrac{16.0244}{142.7494} = 0.1123$

$a = \dfrac{80.9}{16} - 0.1123\left(\dfrac{1368.1}{16}\right) = 5.0563 - 0.1123(85.5063) = 5.0563 - 9.6024 = -4.5461$

c. The change in vital capacity associated with a 1 cm. increase in chest circumference is .1123.

The change in vital capacity associated with a 10 cm. increase in chest circumference is 10(.1123) = 1.123.

d. $\hat{y} = -4.5461 + .1123(85) = 4.9994$

e. No; this is shown by the fact that there are two data points in the data set whose x values are 81.8, but these data points have different y values.

5.33 **a.** $y = 100 + .75(s_y)2 = 100 + 1.5(s_y)$. That person's annual sales would be 1.5 standard deviations above the mean sales of 100.

b. $(y - \bar{y}) = r\dfrac{s_y}{s_x}(x - \bar{x})$, which implies $\dfrac{y - \bar{y}}{s_y} = r\dfrac{x - \bar{x}}{s_x}$. Hence, $1.0 = r(1.5)$ implies

$$r = \frac{-1.0}{-1.5} = .67.$$

5.35 **a.** $\hat{y} = -424.7 + 3.891x$

b. Let $y' = cy$. Then $\bar{y'} = c\bar{y}$.

$b' = \dfrac{\Sigma(x - \bar{x})(cy - c\bar{y})}{\Sigma(x - \bar{x})^2} = \dfrac{c\Sigma(x - \bar{x})(y - \bar{y})}{\Sigma(x - \bar{x})^2} = cb$

$a' = c\bar{y} - cb\bar{x} = c(\bar{y} - b\bar{x}) = ca$

Both the slope and the y intercept are changed by the multiplicative factor c. Thus, the new least squares line is the original least squares line multiplied by c.

Exercises 5.37 – 5.51

5.37 **a.**

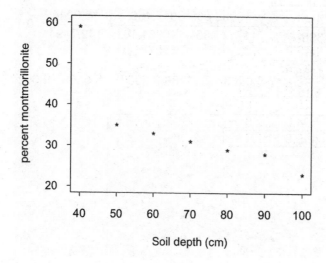

b.

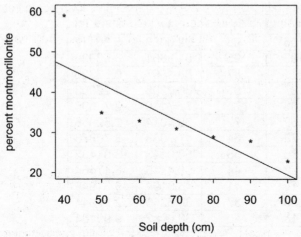

Yes, there appear to be large residuals, those associated with the x-values of 40, 50 and 60.

c.

x	y	$\hat{y}$	$y - \hat{y}$
40	58	46.5	11.5
50	34	42.0	8.0
60	32	37.5	5.5
70	30	33.0	3.0
80	28	28.5	0.5
90	27	24.0	3.0
100	22	19.5	2.5

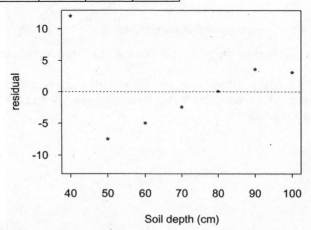

Yes, the residuals for small x-values and large x-values are positive, while the residuals for the middle x-values are negative.

5.39 **a.** The equation of the least-squares line is $\hat{y} = 94.33 - 15.388x$.

b.

x	y	$\hat{y}$	residual
.106	98	92.6989	5.30112
.193	95	91.3601	3.63988
.511	87	86.4667	0.53326
.527	85	86.2205	-1.22053
1.08	75	77.7110	-2.71096
1.62	72	69.4014	2.59856
1.73	64	67.7088	-3.70876
2.36	55	58.0143	-3.01432
2.72	44	52.4746	-8.47464
3.12	41	46.3194	-5.31945
3.88	37	34.6246	2.37544
4.18	40	30.0082	9.99184

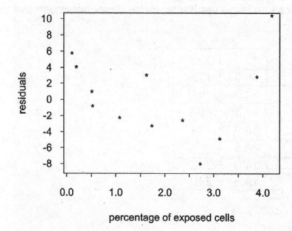

percentage of exposed cells

There appears to be a pattern in the plot. It is like the graph of a quadratic equation.

5.41 **a.** $r^2 = 15.4\%$

b. $r^2 = 16\%$: No, only 16% of the variability in first-year grades can be attributed to an approximate linear relationship between first-year college grades and SAT II score so this does not indicate a good predictor.

54

5.43   **a.** There does appear to be a positive linear relationship between x and y.

b. $\sum x = 798$ $\sum x^2 = 63040$ $\sum y = 643$ $\sum y^2 = 41999$ $\sum xy = 51232$

$$\bar{x} = \frac{798}{15} = 53.2 \qquad \bar{y} = \frac{643}{15} = 42.87$$

$$b = \frac{51232 - \dfrac{(798)(643)}{15}}{63040 - \dfrac{(798)(798)}{15}} = \frac{17024.4}{20586.4} = 0.827$$

$a = 42.87 - 0.827(53.2) = -1.13$
$\hat{y} = -1.13 + 0.827x$

c. For x = 80, $\hat{y} = -1.13 + 0.827(80) = 65.03$

d. SSResid = $\sum y^2 - a\sum y - b\sum xy$
$\qquad\qquad = 41999 - (-1.13)(643) - 0.827(51232)$
$\qquad\qquad = 356.726$

$$s_e = \sqrt{\frac{356.726}{15-2}} = 5.238$$

e.

Rainfall	Runoff	Residual
5	4	0.99344
12	10	1.20463
14	13	2.55068
17	15	2.06976
23	15	-2.89208
30	25	1.31911
40	27	-4.95062
47	46	8.26057
55	38	-6.35522
67	46	-8.2789
72	53	-5.41376
81	70	4.14348
96	82	3.73888
112	99	7.50731
127	100	-3.89728

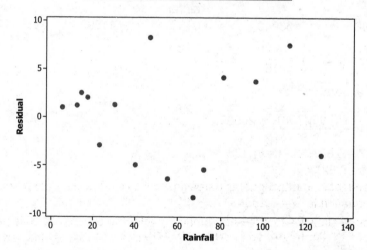

Yes, the variability of the residuals appears to be increasing with x, indicating that a linear relationship may not be appropriate.

b. $r^2 = \dfrac{\left(S_{xy}\right)^2}{(S_{xx})(S_{yy})} = \dfrac{-13.097^2}{(1709.474)(1.012)} = .0992$

c. With r^2 close to 0, the linear relationship between perceived stress and telomere length accounts for a very small proportion of variability in telomere length.

5.45   **a.** $\hat{y} = 766 + .015(9900) = 914.5$

residual = 893   914.5 = 21.5

b. The typical amount that average SAT score deviates from the least squares line is 53.7.

c. Only about 16% of the observed variation in average SAT scores can be attributed to the approximate linear relationship between average SAT scores and expenditure per pupil. The least-squares line does not effectively summarize the relationship between average SAT scores and expenditure per pupil.

5.47   **a.**

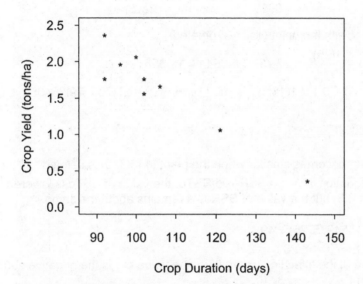

The plot suggests that the least squares line will give fairly accurate predictions. The least squares equation is $\hat{y}$ = 5.20683   .03421x.

b. The summary statistics for the data remaining after the point (143, .3) is deleted are:

$n = 9$ $\Sigma x = 1060$ 143 = 917 $\Sigma x^2 = 114514$ $(143)^2 = 94065$

$\Sigma y = 15.8$ .3 = 15.5 $\Sigma y^2 = 27.82$ $(.3)^2 = 27.73$

$\Sigma xy = 1601.1$ (143)(.3) = 1558.2

$\Sigma x^2 - \dfrac{(\Sigma x)^2}{n} = 94065 - \dfrac{(917)^2}{9} = 94065$ 93432.1111 = 632.8889

$\Sigma xy - \dfrac{(\Sigma x)(\Sigma y)}{n} = 1558.2 - \dfrac{(917)(15.5)}{9} = 1558.2$ 1579.2778 = 21.0778

$$b = \frac{-21.0778}{632.8889} = -.0333$$

$a = 1.7222 - (-.0333)(101.8889) = 1.7222 + 3.3930 = 5.1151$

The least squares equation with the point deleted is $\hat{y} = 5.1151 - .0333x$. The deletion of this point does not greatly affect the equation of the line.

c. For the full data set:

$$SSTo = 27.82 - \frac{(15.8)^2}{10} = 27.82 - 24.964 = 2.856$$

$SSResid = 27.82 - 5.2068338(15.8) - (-.03421541)(1601.1)$
$= 27.82 - 82.2680 + 54.7823 = .3343$

$$r^2 = 1 - \frac{.3343}{2.856} = 1 - .1171 = .8829$$

For the data set with the point (143, .3) deleted:

$$SSTo = 27.73 - \frac{(15.5)^2}{9} = 27.73 - 26.6944 = 1.0356$$

$SSResid = 27.73 - 5.1151(15.5) - (-.0333)(1558.2) = 27.73 - 79.28405 + 51.8881 = .33405$

$$r^2 = 1 - \frac{.33405}{1.0356} = 1 - .3226 = .6774$$

The value of r^2 becomes smaller when the point (143, 0.3) is deleted. The reason for this is that in the equation $r^2 = 1 - SSResid/SSTo$, the value of SSTo is lowered by dropping the point (143, 0.3) but the value of SSResid remains about the same.

5.49 **a.** $\hat{y} = 62.9476 - 0.54975(25) = 49.2$

residual = 49.2 - 70 = -20.8

b. Since the slope of the fitted line is negative, the value of r is the negative square root of r^2. So $r = -\sqrt{r^2} = -\sqrt{0.57} = -0.755$

c. $r^2 = 1 - \dfrac{SSresid}{SSTo}$

$0.57 = 1 - \dfrac{SSresid}{2520}$

Solving for SSResid gives
SSResid = 1083.6

$$s_e = \sqrt{\frac{1083.6}{8}} = 11.64$$

5.51 **a.** When $r = 0$, then $s_e = s_y$. The least squares line in this case is a horizontal line with intercept of $\bar{y}$.

b. When r is close to 1 in absolute value, then s_e will be much smaller than s_y.

c. $s_e = \sqrt{1-(.8)^2}(2.5) = .6(2.5) = 1.5$

d. Letting y denote height at age 6 and x height at age 18, the equation for the least squares line for predicting height at age 6 from height at age 18 is

(height at age 6) $= 46 + .8\left(\dfrac{1.7}{2.5}\right)$[(height at age 18)   70] $= 7.95 + .544$(height at age 18)

The value of s_e is $\sqrt{1-(.8)^2}\,(1.7) = .6(1.7) = 1.02$.

Exercises 5.53 – 5.59

5.53   a.

Fatality Rate vs. Driver's Age

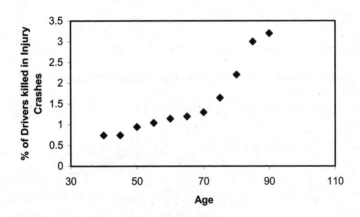

b. Table suggests moving x up or y down so let $y' = \dfrac{1}{y}$.

c.

Age x	Fatality Rate y	y'=1/y
40	0.75	1.33
45	0.75	1.33
50	0.95	1.05
55	1.05	0.95
60	1.15	0.87
65	1.2	0.83
70	1.3	0.77
75	1.65	0.61
80	2.2	0.45
85	3	0.33
90	3.2	0.31

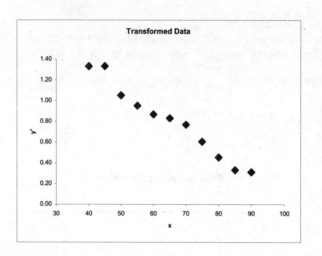

Transformed Data

d. The scatterplot suggests there is a good linear transformation.

e. Using transformed data:

x = 715, $(x)^2 = 49225$ xy' = 516.809,

y' = 8.850, n = 11, $\bar{x} = 65, \bar{y}' = 0.8046$

$$S_{xy'} = \sum xy' - \frac{(\sum x)(\sum y')}{n} = 516.809 - \frac{(715)(8.850)}{11} = -58.441$$

$$S_{xx} = \sum x^2 - \frac{(\sum x)^2}{n} = 49225 - \frac{(715)^2}{11} = 2750$$

The slope, $b = \dfrac{S_{xy'}}{S_{xx}} = \dfrac{-58.441}{2750} = -0.0213$

The intercept, $a = \bar{y} - b\bar{x} = 0.8046 - (-0.0213)(65) = 2.186$

The equation: $\hat{y}' = 2.186 - 0.0213x$ or $\dfrac{1}{\hat{y}} = 2.186 - 0.0213x$, where x = Age and

y = fatality rate. When $x = 78$, $\dfrac{1}{\hat{y}} = 2.186 - 0.0213(78) = .5246$, so $\hat{y} = \dfrac{1}{.5246} = 1.91$.

5.55   **a.** n = 12 $\sum x = 22.4$ $\sum y = 303.1$ $\sum x^2 = 88.58$ $\sum y^2 = 12039.27$ $\sum xy = 241.29$

$$r = \frac{241.29 - \dfrac{(22.4)(303.1)}{12}}{\sqrt{88.58 - \dfrac{(22.4)^2}{12}}\sqrt{12039.27 - \dfrac{(303.1)^2}{12}}} = \frac{-324.50}{\sqrt{46.767}\sqrt{4383.47}}$$

$$= \frac{-324.5}{(6.84)(66.208)} = -0.717$$

b. n=12 $\sum x = 13.5$ $\sum y = 55.74$ $\sum x^2 = 22.441$ $\sum y^2 = 303.3626$
$\sum xy = 47.7283$

$$r = \frac{47.7283 - \dfrac{(13.5)(55.74)}{12}}{\sqrt{22.441 - \dfrac{(13.5)^2}{12}} \sqrt{303.3626 - \dfrac{(55.74)^2}{12}}} = \frac{-14.9792}{\sqrt{7.2535}\sqrt{44.4503}}$$

$$\frac{-14.9792}{(2.693)(6.667)} = -0.835$$

The correlation between $\sqrt{x}$ and $\sqrt{y}$ is .835. Since this correlation is larger in absolute value than the correlation of part **a**, the transformation appears successful in straightening the plot.

5.57   **a.**

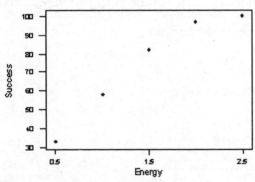

The relationship appears non-linear.

b.   x = 7.5, x^2 = 13.75   xy = 641.05,
y = 370.1, n = 5, $\bar{x} = 1.5, \bar{y} = 74.02$

$$S_{xy} = \sum xy - \frac{(\sum x)(\sum y)}{n} = 641.05 - \frac{(7.5)(370.1)}{5} = 85.9$$

$$S_{xx} = \sum x^2 - \frac{(\sum x)^2}{n} = 13.75 - \frac{(7.5)^2}{5} = 2.5$$

The slope, $b = \dfrac{S_{xy}}{S_{xx}} = \dfrac{85.9}{2.5} = 34.36$

The intercept, $a = \bar{y} - b\bar{x} = 74.02 - 34.36(1.5) = 22.48$
The equation: $\hat{y} = 22.48 + 34.36x$

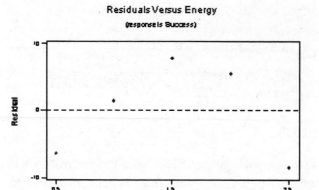

Residuals Versus Energy
(response is Success)

There is a definite curvature in the residual plot confirming the conclusion in part **a.**

c.

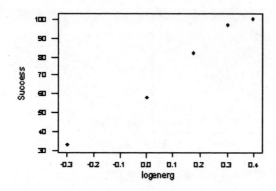

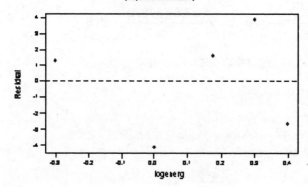

Residuals Versus logenerg
(response is Success)

The value of r^2 is higher and the size of the residuals are smaller for the log transformation.

d. $y = a + b(x')$ where $x' = \log_{10}(x)$

values of x' are: -0.30102, 0, 0.17609, 0.30103, 0.39794

$x' = .5740, \quad (x')^2 = .3706 \quad x'y = 73.2836,$

$y = 370.1, n = 5, \quad \bar{x}' = .1148, \bar{y} = 74.02$

$$S_{xy} = \sum xy - \frac{(\sum x)(\sum y)}{n} = 73.2836 - \frac{(.5740)(370.1)}{5} = 30.796$$

$$S_{xx} = \sum x^2 - \frac{(\sum x)^2}{n} = .3706 - \frac{(.574)^2}{5} = 0.3047$$

The slope, $b = \dfrac{S_{xy}}{S_{xx}} \dfrac{30.796}{0.3047} = 101.07$

The intercept, $a = \bar{y} - b\bar{x} = 74.02 - 101.07(.1148) = 62.417$

The equation: $\hat{y} = 62.417 + 101.07x' \qquad \hat{y} = 62.417 + 101.07 \log(x)$

e. When energy of shock (x) = 1.75, predicted success percent to be 62.417 + 101.07(log 1.75) = 87.0%. When energy of shock is 0.8, the predicted success would be 62.417 + 101.07(log 0.8) = 52.6%

5.59   a.

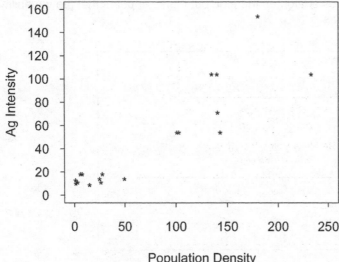

The plot does appear to have a positive slope, so the scatter plot is compatible with the "positive association" statement made in the paper.

b.

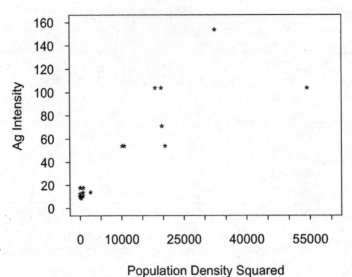

Population Density Squared

This transformation does straighten the plot, but it also appears that the variability of y increases as x increases.

c.

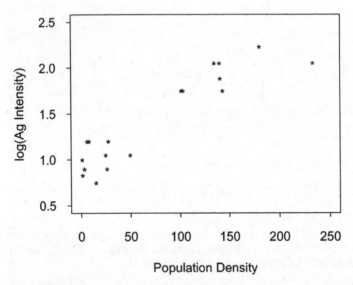

Population Density

The plot appears to be as straight as the plot in b, and has the desirable property that the variability in y appears to be constant regardless of the value of x.

d.

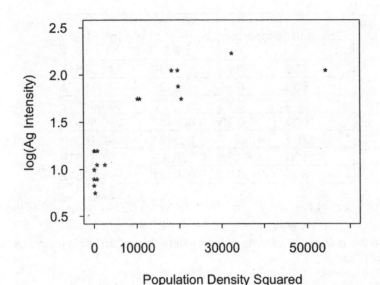

Population Density Squared

This plot has curvature opposite of the plot in part **a**, suggesting that this transformation has taken us too far along the ladder.

Exercises 5.61 – 5.65

5.61 Using x = peak intake and y' = ln(p/1-p): $x = 250$, $x^2 = 16500$ $xy' = 61.75$, $y' = 6.4958$, n = 5, $\bar{x} = 50, \bar{y}' = 1.3$

$$S_{xy'} = \sum xy' - \frac{(\sum x)(\sum y')}{n} = 61.75 - \frac{(250)(6.4958)}{5} = -263.04$$

$$S_{xx} = \sum x^2 - \frac{(\sum x)^2}{n} = 16500 - \frac{(250)^2}{5} = 4000. \text{ The slope,}$$

$$b = \frac{S_{xy'}}{S_{xx}} = \frac{-263.04}{4000} = -0.065876.$$

The intercept, $a = \bar{y}' - b\bar{x} = 1.3 - (-0.06576)(50) = 4.589$

The equation: $\hat{y}' = 4.589 - 0.0659x$

Using the values of a and b from the logistic equation, the probability of survival for a hamster with a peak intake of 40   g: $p = \dfrac{e^{4.589 - 0.0659(40)}}{1 + e^{4.589 - 0.0659(40)}} = .876$

5.63 **a.** It can be seen form the table that as the elevation increases, the Lichen becomes less common.

b.

Elevation	Proportion	$\dfrac{p}{(1-p)}$	$y' = \ln\left(\dfrac{p}{1-p}\right)$
400	0.99	99	4.59512
600	0.96	24	3.178054
800	0.75	3	1.098612
1000	0.29	0.408451	-0.89538
1200	0.077	0.083424	-2.48382
1400	0.035	0.036269	-3.31678
1600	0.01	0.010101	-4.59512

The resulting best fit line is: $y' = a + bx = 7.652 - 0.0079x$, where y is the proportion of plots with lichen and x = elevation.

c. To estimate the proportion of plots of land where the lichen is classified as "common" at an elevation of 900m:

$$p = \frac{e^{7.652-0.0079(900)}}{1 + e^{7.652-0.0079(900)}} = .632$$

5.65   **a.**

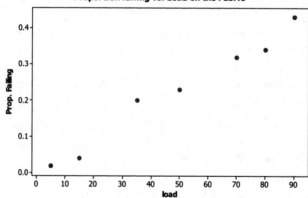

Proportion failing vs. Load on the Fabric

b.

Load	Prop. failing	$\dfrac{p}{(1-p)}$	$y' = \ln\left(\dfrac{p}{1-p}\right)$
5	0.02	0.020408	-3.89182
15	0.04	0.041667	-3.17805
35	0.2	0.25	-1.38629
50	0.23	0.298701	-1.20831
70	0.32	0.470588	-0.75377
80	0.34	0.515152	-0.66329
90	0.43	0.754386	-0.28185

The resulting best fit line is: $y' = a + bx = -3.579 + 0.0397x$, where y is the proportion of fabrics failing and $x =$ the load applied.

The positive slope, $b > 0$, shows that as the load or forces increases, the proportion of the fabrics that fail also increases.

c. When the load is 60, $p = \dfrac{e^{-3.579+0.0397(60)}}{1+e^{-3.579+0.0397(60)}} = .232$.   .232 lbs per sq. in.

d. When the failure rate is 5%, p = 0.05, so: $\ln\left(\dfrac{p}{(1-p)}\right) = a + bx$

$$\ln\left(\frac{0.05}{(1-0.05)}\right) = -3.579 + 0.0397x$$

$$x = \frac{\ln(0.0526)+3.579}{0.0397} = 15.98 .$$

To have less than a 5% chance of a wardrobe malfunction, a maximum force of 15.5 lbs/sq in. might be suggested.

Exercises 5.67 – 5.77

5.67   **a.** For this data set, n= 13, $\sum x = 5.92$, $\sum x^2 = 3.8114$, $\sum y = 10.47$,

$$\sum y^2 = 9.885699, \quad \sum xy = 5.8464 , \quad \bar{x} = 0.455, \quad \bar{y} = 0.805$$

$$b = \frac{5.8464 - \dfrac{(5.92)(10.47)}{13}}{3.8114 - \dfrac{(5.92)(5.92)}{13}} = \frac{5.8464 - 4.7679}{3.8114 - 2.6959} = \frac{1.0785}{1.1155} = 0.9668$$

a = 0.805 – 0.9668(0.455) = 0.3651
The least squares regression line is $\hat{y} = 0.3651 + 0.9668x$

b. For a value of x = 0.5, $\hat{y} = 0.3651 + 0.9668(0.5) = 0.8485$

5.69 $r^2 = 1 - \dfrac{5987.16}{17409.60} = 1 - .3439 = .6561$

So 65.61% of the observed variation in age is explained by a linear relationship between percentage of root with transparent dentine for premolars and age.

$$s_e^2 = \frac{5987.16}{36-2} = \frac{5987.16}{34} = 176.0929$$

$$s_e = \sqrt{176.0929} = 13.27$$

The typical amount by which an observed age deviates from the least squares line of percentage of root with transparent dentine and age is 13.27.

5.71   **a.** $\sum x^2 - \frac{(\sum x)^2}{n} = 62.600235 - \frac{(22.027)^2}{12} = 62.600235 - 40.43239 = 22.16784$

$\sum xy - \frac{(\sum x \sum y)}{n} = 1114.5 - \frac{(22.027)(793)}{12} = 1114.5 \quad 1455.61758 = \quad 341.11758$

$b = \frac{-341.11758}{22.16784} = -15.38795$

$a = 66.08333$ (15.38795)(1.83558) = 66.08333 + 28.24586 = 94.32919

The least squares equation is $\hat{y} = 94.33$ 15.388x

b. SSTo $= \sum y^2 - \frac{(\sum y)^2}{n} = 57939 - \frac{(793)^2}{12} = 57939$ 52404.08333 $= 5534.91667$

SSResid $= 57939$ 94.32919(793)   (15.38795)(1114.5)

$= 57939$ 74803.04767 + 17149.87028 = 285.82261

c. $r^2 = 1 - \frac{285.82261}{5534.91667} = 1 - .05164 = .94836$ or 94.836%

d. $s_e^2 = \frac{285.82261}{10} = 28.582261$

$s_e = \sqrt{28.582261} = 5.34624$

A typical prediction error would be about 5.35 percent.

e. Since the slope of the fitted line is negative, the value of r is the negative square root of r^2.

So $r = -\sqrt{r^2} = -\sqrt{.94836} = -.97384$.

5.73   **a.** A value of .11 for r indicates a weak linear relationship between annual raises and teaching evaluations.

b. $r^2 = (.11)^2 = .0121$

5.75   The summary values are: n = 13, $\sum x = 91$, $\sum y = 470$, $\sum x^2 = 819$, $\sum y^2 = 19118$

$\sum xy = 3867$

$\sum xy - \frac{(\sum x)(\sum y)}{n} = 577, \sum x^2 - \frac{(\sum x)^2}{n} = 182, \quad \sum y^2 - \frac{(\sum y)^2}{n} = 2125.6923$

a. $b = \frac{577}{182} = 3.1703$ a = 36.1538   3.1703(7) = 13.9617

The equation of the estimated regression line is $\hat{y} = 13.9617 + 3.1703x$

b.

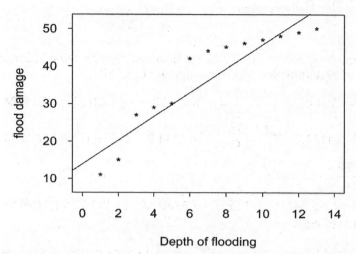

The plot with the line drawn in suggests that perhaps a simple linear regression model may not be appropriate. The scatterplot suggests that a curvilinear relationship may exist between flood depth and damage. The points for small x-values or large x-values are below the line, while points for x-values in the middle range are above the line.

c. When x = 6.5, $\hat{y}$ = 13.9617 + 3.1703(6.5) = 34.5687

d. The scatterplot in part **b** suggests that the value of damage levels off at between 45 and 50 when the depth of flooding is in excess of 10 feet. Using the least squares line to predict flood damage when x = 18 would yield a very high value for damage and result in a predicted value far in excess of actual damage. Since x = 18 is outside of the range of x-values for which data has been collected, we have no information concerning the relationship in the vicinity of x = 18. All of these reasons suggest that one would not want to use the least squares line to predict flood damage when depth of flooding is 18 feet.

5.77   **a.**

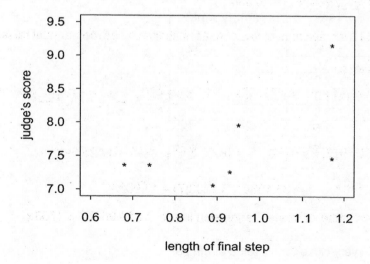

b. $\sum x^2 - \dfrac{(\sum x)^2}{n} = .2157$, $\sum y^2 - \dfrac{(\sum y)^2}{n} = 3.08$, $\sum xy - \dfrac{(\sum x)(\sum y)}{n} = 0.474$

$$b = \dfrac{.474}{.2157} = 2.1975$$

$a = 7.6 \quad (2.1975)(.93286) = 5.550$

The least squares line is $\hat{y} = 5.550 + 2.1975\,x$

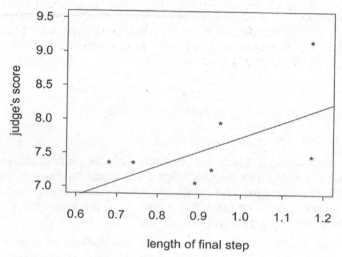

length of final step

c. $s_x = \sqrt{.2157/6} = .1896$, $s_y = \sqrt{3.08/6} = .7165$

$$r = \dfrac{.474}{6(.1896)(.7165)} = .5815$$

This value of r suggests a moderate positive linear relationship between x and y.

d.

x rank	y rank	(x rank)(y rank)
6.5	5.0	32.5
6.5	7.0	45.5
4.0	2.0	8.0
3.0	1.0	3.0
1.0	3.5	3.5
2.0	3.5	7.0
5.0	6.0	30.0
		129.5

$$r_s = \dfrac{129.5 - \dfrac{7(8)^2}{4}}{\dfrac{7(6)(8)}{12}} = \dfrac{17.5}{28} = .625$$

This value is very close to the value of r in part **c.**

Chapter 6

Exercises 6.1 – 6.11

6.1 A chance experiment is any activity or situation in which there is uncertainty about which of two or more possible outcomes will result.
Consider tossing a coin two times and observing the outcome of the two tosses. The possible outcomes are (H, H), (H, T), (T, H), and (T, T), where (H, H) means both tosses resulted in Heads, (H, T) means the first toss resulted in a Head and the second toss resulted in a Tail, etc. This is an example of a chance experiment with four possible outcomes.

6.3 **a.** Sample space = {(A, A), (A, M), (M, A), (M, M)}.
b.

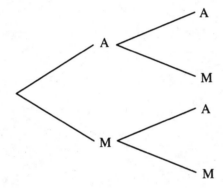

c. B = {(A, A), (A, M), (M, A)}
C = {(A, M), (M, A)}
D = {(M, M)}
Only D is a simple event.
d. B and C = {(A, M), (M, A)} = C
B or C = {(A, A), (A, M), (M, A)} = B

6.5 **a.**

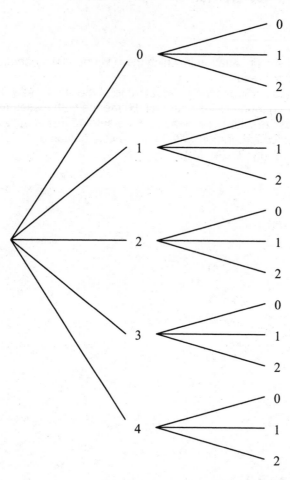

# of defective tires   # of defective headlights

b. A^c = {(0,2), (1,2), (2,2), (3,2), (4,2)} where (i,j) means that the number of defective tires is i and the number of defective headlights is j.

A $\cup$ B = {(0,0), (1,0), (2,0), (3,0), (4,0), (0,1), (1,1), (2,1), (3,1), (4,1), (0,2), (1,2)} where (i,j)

means that the number of defective tires is i and the number of defective headlights is j.

A $\cap$ B = {(0,0), (1,0), (0,1), (1,1)}

c. C = {(4,0), (4,1), (4,2)}. A and C are not disjoint because the outcomes (4,0) and (4,1) are in both A and C. B and C are disjoint because the outcomes in B are of the form (i,j) where i =0 or 1 but the outcomes in C are of the form (i,j) where i=4. Recall that the notation (i,j) means that the number of defective tires is i and the number of defective headlights is j.

6.7 **a.**

First book selected   Second book selected   Third book selected

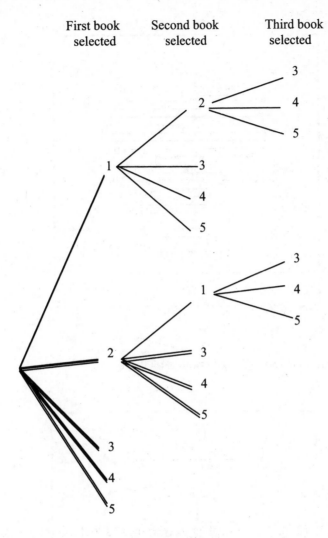

b. Exactly one book is examined when the first selected book is copy 3, 4, or 5. So A = {3,4,5}.
c. C = {5, (1,5), (2,5), (1,2,5), (2,1,5)}. The notation used is as follows. The outcome 5 means that copy 5 is selected first. The outcome (1,5) means that copy 1 is selected first and copy 5 is selected second, and so on.

6.9 **a.** A = {NN, DNN, NDN}
b. B = {DDNN, DNDN, NDDN}
c. The number of possible outcomes is infinite.

a.

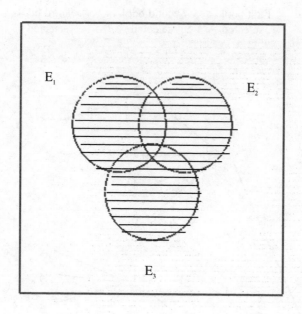

b. The required region is the area common to all three circles.

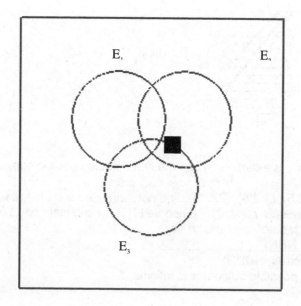

c.

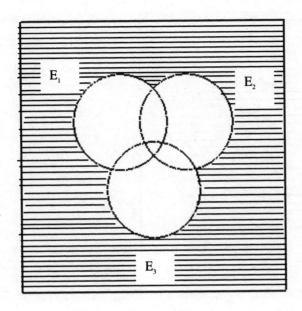

d.

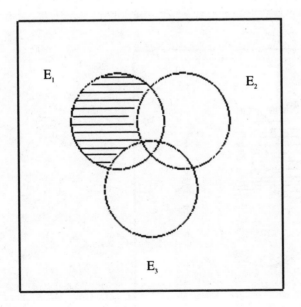

e.

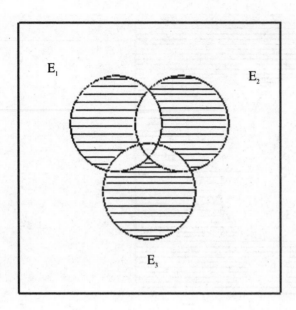

f.

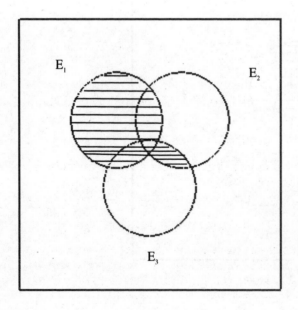

Exercises 6.13 – 6.27

6.13   **a.** A = {(C,N), (N,C), (N,N)}. So P(A) = 0.09 + 0.09 + 0.01 = 0.19.
  b. B = {(C,C), (N,N)}. So P(B) = 0.81 + 0.01 = 0.82.

6.15   **a.** In the long run, 1% of all people who suffer cardiac arrest in New York City survive.
  b. 1% of 2329 is .01(2329) = 23.29. So in this study there must have been only 23 or 24 survivors.

6.17   **a.** P(twins) = $\dfrac{500,000}{42,005,100}$ = .0119

  b. P(quadruplets) = $\dfrac{100}{42,005,100}$ = .00000238

  c. P(more than a single child) = $\dfrac{505,100}{42,005,100}$ = .012

6.19   **a.** P(owns shares in balanced fund) = 0.07.
  b. P(owns shares in a bond fund) = 0.15 + 0.10 + 0.05 = 0.30.
  c. P(does not own shares in a stock fund) = 1 - P(owns shares in a stock fund)
    = 1 - (0.18 + 0.25) = 0.57.

6.21   No, it is not true. The addition rule is for disjoint events. The events "midsize" and "4 3/8 inch grip" are not disjoint since P(midsize AND 4 3/8 inch grip) = 0.10 is greater than zero.

6.23   **a.** The 24 outcomes (including the outcome (2, 4, 3, 1)) are
    (1, 2, 3, 4), (1, 2, 4, 3), (1, 3, 2, 4), (1, 3, 4, 2), (1, 4, 2, 3), (1, 4, 3, 2)
    (2, 1, 3, 4), (2, 1, 4, 3), (2, 3, 1, 4), (2, 3, 4, 1), (2, 4, 1, 3), (2, 4, 3, 1)
    (3, 1, 2, 4), (3, 1, 4, 2), (3, 2, 1, 4), (3, 2, 4, 1), (3, 4, 1, 2), (3, 4, 2, 1)
    (4, 1, 2, 3), (4, 1, 3, 2), (4, 2, 1, 3), (4, 2, 3, 1), (4, 3, 1, 2), (4, 3, 2, 1)
  b. The required outcomes are
    (1, 2, 4, 3), (1, 4, 3, 2), (1, 3, 2, 4), (4, 2, 3, 1), (3, 2, 1, 4), (2, 1, 3, 4).

    P(exactly two of the 4 books will be returned to the correct owner) = $\dfrac{6}{24}$ = 0.25 .

  c. The outcomes corresponding to the event "exactly one students receives his/her book" are
    (1, 3, 4, 2), (1, 4, 2, 3), (2, 3, 1, 4), (2, 4, 3, 1), (3, 1, 2, 4), (3, 2, 1, 4), (4, 1, 3, 2), (4, 2, 1, 3).

    Hence the probability of this event = $\dfrac{8}{24}$ = 0.33333.

  d. If three students receive their books, then the fourth student must receive his/her book also. So, the required probability is 0.

  e. The outcomes corresponding to this event are (1, 2, 3, 4), (1, 2, 4, 3), (1, 3, 2, 4), (1, 4, 3, 2), (2, 1, 3, 4), (3, 2, 1, 4), (4, 2, 3, 1). Hence the required probability is $\dfrac{7}{24}$ = 0.291667.

6.25   **a.** Suppose the students selected are from Mathematics and Physics. We abbreviate this outcome by {M, P}. With this notation, the ten possible outcomes are {B, C}, {B, M}, {B, P}, {B, S}, {C, M}, {C, P}, {C, S}, {M, P}, {M, S}, {P, S}.
  b. Probability of each simple event = 1/10 = 0.1.

c. P(one member is from Statistics) = 4/10 = 0.4.

d. Assuming that the laboratory sciences are Biology, Chemistry, and Physics, the required probability is 3/10 = 0.3.

6.27 **a.** $P(O_1) + P(O_2) + P(O_3) + P(O_4) + P(O_5) + P(O_6) = 1$ implies that $p + 2p + p + 2p + p + 2p = 1$.

So. 9p = 1 which yields p = 1/9. The probabilities of the six simple events are as follows:
$$P(O_1) = P(O_3) = P(O_5) = 1/9$$
$$P(O_2) = P(O_4) = P(O_6) = 2/9.$$

b. P(odd number) = P(1 or 3 or 5) = 3(1/9) = 3/9 = 1/3 = 0.3333.

P(at most 3) = P(1 or 2 or 3) = 1/9 + 2/9 + 1/9 = 4/9 = 0.4444.

c. Now, $P(O_1) + P(O_2) + P(O_3) + P(O_4) + P(O_5) + P(O_6) = 1$ implies that

c + 2c + 3c + 4c + 5c + 6c = 1, from which we get c = 1/21.

Hence $P(O_1) = 1/21$; $P(O_2) = 2/21$; $P(O_3) = 3/21$; $P(O_4) = 4/21$; $P(O_5) = 5/21$; $P(O_6) = 6/21$.

P(odd number) = 1/21 + 3/21 + 5/21 = 9/21 = 0.428571.

P(at most 3) = 1/21 + 2/21 + 3/21 = 6/21 = 0.285714.

Exercises 6.29 – 6.39

6.29 **a.** Let E = the event that critic 1 gave a "thumbs up" to the chosen movie and F = the event that critic 2 gave a "thumbs up" to the selected movie. Then $P(E) = 15/60 = \frac{1}{4}$, $P(F) = 20/60 = 1/3$, and $P(E \cap F) = 10/60 = 1/6$.

So, the required probability = $P(F \mid E) = \dfrac{P(E \cap F)}{P(E)} = \dfrac{1/6}{1/4} = 0.6667$.

b. Using the given information we construct the following table showing the various possible outcomes and their frequencies of occurrence.

	Critic 2 "thumbs up"	Critic 2 "thumbs down"	Total
Critic 1"thumbs up"	10	5	15
Critic 1"thumbs down"	10	35	45
Total	20	40	60

P(thumbs down from critic 1|thumbs down from critic 2) $= \dfrac{35}{40} = 0.875$.

6.31 The statement in the article implies the following two conditions:

(1) $P(D \mid Y^c) > P(D \mid Y)$ and

(2) $P(Y) > P(Y^c)$.

This claim is consistent with the information given in I but not with any of the others. In II and III, condition (1) is violated; in IV, condition (2) is violated; in V and VI, both conditions are violated.

6.33 **a.** $P(F \mid F^P) = \dfrac{432}{562} = 0.769$

b. $P(M|M^P) = \dfrac{390}{438} = 0.890$

c. Ultrasound appears to be more reliable in predicting boys

6.35   **a. i.**   Required conditional probability = $\dfrac{135}{182} = 0.74176$.

   ii.   Required conditional probability = $\dfrac{173 + 206}{210 + 231} = 0.8594$.

b.

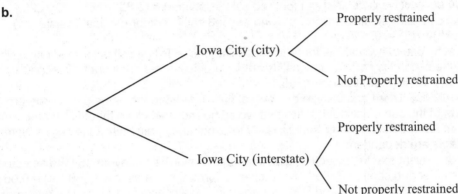

6.37   **a.** $0.1 + 0.175 = 0.275$

   b. $\dfrac{0.1}{0.5} = 0.2$

   c. $\dfrac{0.325}{0.5} = 0.65$

   d. $\dfrac{0.325}{0.725} = 0.448$

   e. The probabilities in **c** and **d** are not equal because the events "wearing a seatbelt" and "being female" are not independent. It is clearly more likely an adult will regularly use a seatbelt if they are female than if they are male.

6.39   **a. i.** $P(S) = \dfrac{456}{600} = 0.76$ **ii.** $P(S|A) = \dfrac{215}{300} = 0.717$ **iii.** $P(S|B) = \dfrac{241}{300} = 0.803$

   iv. Drug B appears to have a higher survival rate

   b. i. $P(S) = \dfrac{140}{240} = 0.583$ **ii.** $P(S|A) = \dfrac{120}{200} = 0.600$ **iii.** $P(S|B) = \dfrac{20}{40} = 0.5$

   iv. Drug A appears to have the higher survival rate.

   c. i. $P(S) = \dfrac{316}{360} = 0.878$ **ii.** $P(S|A) = \dfrac{95}{100} = 0.95$ **iii.** $P(S|B) = \dfrac{221}{260} = 0.85$

   iv. Drug A appears to have the higher survival rate

d. Women seem to respond well to both treatments. Men seem to respond very well to treatment A but not so well to treatment B. When the data is combined, the small quantity of data collected for men on treatment B - only 20% of those men who had treatment A, becomes "lost" in the rest of the data.

Exercises 6.41 – 6.57

6.41 **a.** About 85% of all the past calls were for medical assistance.

b. P(call is not for medical assistance) = $1 - 0.85 = 0.15$.

c. P(two successive calls are both for medical assistance) = $(0.85)(0.85) = 0.7225$.

d. P(first call is for medical assistance and second call is not for medical assistance) = $(0.85)(0.15) = 0.1275$.

e. P(exactly one of two calls is for medical assistance) = P(first call is for medical assistance and the second is not) + P(first call is not for medical assistance but the second is) = $(0.85)(0.15) + (0.15)(0.85) = 0.255$.

f. Probably not. There are likely to be several calls related to the same event—several reports of the same accident or fire that would be received close together in time.

6.43 No, a randomly selected adult is more likely to experience pain daily if they are a woman. The events are dependent.

6.45 Let A = the event that the selected adult is a female and B = the event that the selected adult favors stricter gun control. Based on the given information we estimate P(B | A) = 0.66 and P(B) = 0.56. Since P(B | A) and P(B) are not equal, we conclude that the two events A and B are dependent outcomes.

6.47 P(jury duty call two years in a row) = $(0.15)(0.15) = 0.0225$.
P(jury duty call three years in a row) = $(0.15)(0.15)(0.15) = 0.003375$.

6.49 **a.** P(none of the 10 calls lead to a reservation) = $(0.7)^{10} = 0.0282475249$.

b. It was assumed that the outcomes of the different calls are independent, i.e., whether one call resulted in a reservation or not has no influence on the results of other calls.

c. P(at least one call leads to a reservation) = $1 - 0.0282475249 = 0.9717524751$.

6.51 **a.** P(1-2 subsystem works) = P(1 works) P(2 works) = $(0.9)(0.9) = 0.81$.

b. P(1-2 subsystem does not work) = $1 - $ P(1-2 subsystem works) = $1 - 0.81 = 0.19$.
P(3-4 subsystem does not work) = $1 - $ P(3-4 subsystem works) = $1 - (0.9)(0.9) = 0.19$.

c. Using the results of part **b**, P(system won't work) = P(1-2 subsystem doesn't work & 3-4 subsystem doesn't work) = P(1-2 subsystem doesn't work) P(3-4 subsystem doesn't work) = $(0.19)(0.19) = 0.0361$.
So P(system will work) = $1 - $ P(system won't work) = $1 - 0.0361 = 0.9639$.

d. If a 5-6 subsystem were added, then P(system will work) = $1 - $ P(system won't work) = $1 - (0.19)(0.19)(0.19) = 0.993141$.

e. If there were 3 components in series in each subsystem, then P(system will work) = $1 - $ P(system won't work) = $1 - $ P(first subsystem doesn't work) P(second subsystem doesn't work). The probability that the first subsystem doesn't work = $1 - (0.9)(0.9)(0.9) = 1 - 0.729 = 0.271$. This is also the probability that the second subsystem won't work. So P(system works) = $1 - (0.271)(0.271) = 0.926559$.

6.53 a. The expert assumed that a valve stem could be in any one of the "clock positions" with equal probability, so that there is a 1 in 12 chance for each valve to be in any given clock hour position. In addition, the expert assumed that the positions of the two valve stems are independent. Hence, the expert computed the probability of a "match" on both valves to be the product $(1/12)(1/12) = 1/144$.

b. The assumption of "independence" of the two valve stems is highly questionable because the valve stems turn in a "synchronized" manner. This would make the probability of the valve stems being in their original positions greater than the $1/144$ value computed by the expert.

6.55 a. If the first board is defective, then the probability that the second board is defective = $39/4999$. If the first board is not defective, then the probability that the second board is defective = $40/4999$. Since the probability that the second board is defective changes depending on whether or not the first board is defective, the events E_1 and E_2 are dependent.

b. $P(\text{not } E_1) = 1 - P(E_1) = 1 - (40/5000) = 0.992$.

c. $P(E_2 \mid E_1) = 39/4999$ and $P(E_2 \mid \text{not } E_1) = 40/4999$. They are nearly equal.

d. Yes.

6.57 a. $P(CC) = (50/800)(50/800) = 0.00391$.

b. $P(CC) = P(\text{first answer is correct}) \, P(\text{second answer is correct} \mid \text{first answer is correct})$
= $(50/800)(49/799) = 0.00383$. This is very close to the probability computed in part **a**.

Exercises 6.59 – 6.73

6.59 a. $P(E) = 6/10 = 0.6$

b. $P(F \mid E) = 5/9 = 0.5556$

c. $P(E \text{ and } F) = P(E) \, P(F|E) = (6/10)(5/9) = 30/90 = 1/3$

6.61 a. $P(\text{individual has to stop at at least one light}) = P(E \cup F) = P(E) + P(F) - P(E \cap F)$
= $0.4 + 0.3 - 0.15 = 0.55$.

b. $P(\text{individual doesn't have to stop at either traffic light}) = 1 - P(\text{must stop at at least one light})$
= $1 - 0.55 = 0.45$ (using the result of part **a**).

c. $P(\text{stop at exactly one of the two lights}) = P(\text{stop at least one of the two lights}) - P(\text{stop at both lights}) = 0.55 - 0.15 = 0.40$.

d. $P(E \text{ only}) = P(E \text{ and not } F) = P(E) - P(E \cap F) = 0.4 - 0.15 = 0.25$.

6.63 The likelihood of using a cell phone is different for of each type of vehicle; for instance if you drive a van or SUV, you are more likely to use a cell phone than if you drive a pick-up truck.

6.65 **a. i.** 0.5   **ii.** 0.5   **iii.** 0.99   **iv.** .01

b. $P(TD) = P(TD \cap D) + P(TD \cap C) = P(TD|D)P(D) + P(TD|C)P(C) =$
$= (.99)(.01) + (.99)(.01) = 2(.99)(.01) = .0198$

c. $P(C|TD) = \dfrac{P(C)P(TD \mid P(C))}{P(TD \mid P(C)) + P(TD) \mid P(D)} = \dfrac{(.99)(.01)}{2(.99)(.01)} = 0.5$

Yes, it is consistent with the argument given in the quote.

6.67 **a.** In 4 consecutive years, there are 3 years of 365 days (1095 non leap days) and 1 year of 366 days (365 non leap days and one leap day) – a total of 1461 days, of which one, Feb 29th is a leap day. Hence a leap day occurs once in 1461 days.

b. If babies are induced or born by C-section, they are less likely to be born at weekends or on holidays so they are not equally likely to be born on the 1461 days.

c. 1 in 2.1 million is the probability of picking a mother and baby randomly and finding *both* to be a leap-day mom-baby ($1/1461^2$). The probability of a leap year baby becoming a leap year mom means that the mom's birthday is already fixed, so the hospital spokesperson's probability is too small.

6.69 **a.** Assume that the results of successive tests on the same individual are independent of one another. Let F_1 = the event that the selected employee tests positive on the first test and F_2 = the event that he/she tests positive on the second test. Then, P(employee uses drugs and test positive twice) = $P(E)P(F_1|E)P(F_2|E) = (0.1)(0.90)(0.90) = 0.081$.

b. P(employee tests positive twice)
  = P(employee uses drugs and tests positive twice)
   + P(employee doesn't use drugs and tests positive twice)
  = $P(E) P(F_1|E)P(F_2|E) + P(E^c) P(F_1|E^c)P(F_2|E^c)$
  = $(0.1)(0.90)(0.90) + (0.9)(0.05)(0.05) = 0.081 + 0.00225 = 0.08325$.

c. P(uses drugs | tested positive twice) =
$\dfrac{P(\text{uses drugs and tests positive twice})}{P(\text{tests positive twice})} = \dfrac{0.081}{0.08325} = 0.97297$.

d. P(tests negative on the first test OR tests positive on the first test but tests negative on the second test | does use drugs) = $P(F_1^c \mid E) + P(F_1 \text{ and } F_2^c \mid E)$
  = $[1 - P(F_1 \mid E)] + P(F_1 \mid E)[1 - P(F_2 \mid E)]$
  = $[1 - 0.9] + (0.9)(1 - 0.9) = 0.1 + 0.09 = 0.19$
  (using $P(F_1 \mid E) = 0.90$, from part **a** of Problem **6.68**).

e. The benefit of using a retest is that it reduces the rate of false positives from 0.05 to 0.02703 (obtained as $1 - 0.97297$, using the result from part **c**), but the disadvantage is that the rate of false negatives has increased from $P(F^c \mid E) = 0.10$ to 0.19 (see part **d** above). Retests also involve additional expenses.

6.71 Based on the given information we can construct the following table.

	Basic Model	Deluxe Model	Total
Buy extended warranty	12%	30%	42%
Do not buy extended warranty	28%	30%	58%
Total	40%	60%	100%

So P(Basic model | extended warranty) = 12/42 = 0.285714.

6.73   Let A = the event that he takes the small car; B = the event that he takes the big car; C = the event that he is on time. Then, based on the given information we have P(A) = 0.75, P(B) = 0.25, P(C|A) = 0.9, and P(C|B) = 0.6. We need P(A|C). Using Bayes' Rule we get

$$P(A|C) = \frac{P(C|A)P(A)}{P(C|A)P(A) + P(C|B)P(B)} = \frac{(0.9)(0.75)}{(0.9)(0.75) + (0.6)(0.25)} = 0.8182.$$

Exercises 6.75 – 6.83

6.75   **a.** P(on time delivery in Los Angeles) = 425/500 = 0.85.
   b. P(late delivery in Washington, D.C) = (500 – 405)/500 = 95/500 = 0.19.
   c. Assuming that whether or not one letter is delivered late is "independent" of the on-time delivery status of any other letter, we calculate P(both letters delivered on-time in New York) = (415/500)(415/500) = 0.6889. (The independence assumption may not be a valid assumption).
   d. P(on-time delivery nationwide) = 5220/6000 = 0.87.

6.77   **a.** P(male) = (200+800+1500+1500+900+1500+300)/17000 = 6700/17000 = 0.3941.
   b. P(agriculture) = 3000/17000 = 0.1765.
   c. P(male AND from agriculture) = 900/17000 = 0.0529.
   d. P(male AND not from agriculture) = 5800/17000 = 0.3412.

6.79   The results will vary from one simulation to another. The approximate probabilities given were obtained from a simulation done by computer with 100,000 trials.
   a. 0.71293   **b.** 0.02201

6.81   The simulation results will vary from one simulation to another. The approximate probability should be around .8504.

6.83   **a.** The simulation results will vary from one simulation to another. The approximate probability should be around 0.6504.
   b. The decrease in the probability of on time completion for Jacob made the biggest change in the probability that the project is completed on time.

Exercises 6.85 – 6.97

6.85   P(parcel went via A_1 and was late) = P(went via A_1)P(late | went via A_1) = (0.4)(0.02) = 0.008.

6.87   **a.** $(.3)^5 = .00243$

   b. $(.3)^5 + (.2)^5 = .00243 + .00032 = .00275$

   c. Select a random digit. If it is a 0, 1, or 2, then award A one point. If it is a 3 or 4, then award B one point. If it is 5, 6, 7, 8, or 9, then award A and B one-half point each. Repeat the selection process until a winner is determined or the championship ends in a draw (5 points each). Replicate this entire procedure a large number of times. The estimate of the probability that A wins the championship would be the ratio of the number of times A wins to the total number of replications.
   d. It would take longer if no points for a draw are awarded. It is possible that a number of games would end in a draw and so more games would have to be played in order for a player to earn 5 points.

6.89 P(need to examine at least 2 disks) = 1 – P(get a blank disk in the first selection) = 1 – 15/25 = 0.4.

6.91 Let the five faculty members be A, B, C, D, E with teaching experience of 3, 6, 7, 10, and 14 years respectively. The two selected individuals will have a total of at least 15 years of teaching experience if the selected individuals are A and E, or B and D, or B and E, or C and D, or C and E, or D and E. Since there are a total of 10 equally likely choices and 6 out of these meet the specified requirement, P(two selected members will have a total of at least 15 years experience) = 6/10 = 0.6.

6.93 Total number of viewers = 2517.

a. P(saw a PG movie) = $\dfrac{179 + 87}{2517}$ = 0.1057.

b. P(saw a PG or a PG-13 movie) = $\dfrac{420 + 323 + 179 + 114 + 87}{2517}$ = 0.4462.

c. P(did not see an R movie) = $\dfrac{2517 - 600 - 196 - 205 - 139}{2517}$ = 0.5471.

6.95 **a.** P(neither is selected for testing | batch has 2 defectives) = (8/10)(7/9) = 56/90 = 0.6222.

b. P(batch has 2 defectives and neither selected for testing) = P(batch has 2 defectives) P(neither is selected for testing | batch has 2 defectives) = (0.20)(0.6222) = 0.1244.

c. P(neither component selected is defective) = P(batch has 0 defectives & neither selected component is defective) + P(batch has 1 defective & neither selected component is defective) + P(batch has 2 defectives & neither selected component is defective) = (0.5)(1) + (0.3)(9/10)(8/9) + (0.2)(8/10)(7/9) = 0.5 + 0.24 + 0.1244 = 0.8644.

6.97 $P(E \cap F \cap G \cap H) = P(E)P(F|E)P(G|E \cap F)P(H|E \cap F \cap G)$ = (20/25)(19/24)(18/23)(17/22) = 0.3830. P(at least one bad bulb) = 1 – P(all 4 bulbs are good) = 1 – 0.383 = 0.617.

Chapter 7

Exercises 7.1 – 7.7

7.1   **a.** discrete
      b. continuous
      c. discrete
      d. discrete
      e. continuous

7.3   The possible y values are the set of all positive integers.
      Some possible outcomes are LS, RRS, S, LRRRLLRLLS, and LLLLS, with corresponding y values equal to 2, 3, 1, 10, and 5, respectively.

7.5   Possible values of y (in feet) are the real numbers in the interval $0 \leq y \leq 100$. The variable y is a continuous variable.

7.7   **a.** Possible values for x are 3, 4, 5, 6, 7.
      b. If y = first number – second number, then possible values of y are –3, – 2, – 1, 1, 2, and 3.
      c. Possible values of z are 0, 1, 2.
      d. Possible values of w are 0, 1.

Exercises 7.9 – 7.19

7.9   **a.** $p(4) = 1 - p(0) - p(1) - p(2) - p(3) = 1 - 0.65 - 0.20 - 0.10 - 0.04 = 0.01$.
      b. In the long run, the proportion of cartons that have exactly one broken egg will equal 0.20.
      c. $P(y \leq 2) = 0.65 + 0.20 + 0.10 = 0.95$. This means that, in the long run, the proportion of cartons that have two or fewer broken eggs will equal 0.95.
      d. $P(y < 2) = 0.65 + 0.20 = 0.85$. This probability is less than the probability in part **c** because the event y = 2 is now not included.
      e. P(exactly 10 unbroken eggs) = P(exactly 2 broken eggs) = $P(y = 2) = 0.10$.
      f. P(at least 10 unbroken eggs) = P(0 or 1 or 2 broken eggs) = $P(y \leq 2) = 0.95$ (from part **c**).

7.11   **a.** P(airline can accommodate everyone who shows up)
      $= P(x \leq 100) = 0.05 + 0.10 + 0.12 + 0.14 + 0.24 + 0.17 = 0.82$.
      b. P(not all passengers can be accommodated) = $P(x > 100) = 1 - P(x \leq 100) = 1 - 0.82 = 0.18$.
      c. P(number 1 standby will be able to take the flight) = $P(x \leq 99)$
      $= 0.05 + 0.10 + 0.12 + 0.14 + 0.24 = 0.65$.
      P(number 3 standby will be able to take the flight) = $P(x \leq 97) = 0.05 + 0.10 + 0.12 = 0.27$.

7.13 Results will vary. One particular set of 50 simulations gave the following results.

Value of x	Frequency	Relative Frequency
0	13	13/50 = 0.26
1	32	32/50 = 0.64
2	5	5/50 = 0.1

These relative frequencies are quite close to the theoretical probabilities given in part **b** of Problem 7.12.

7.15 **a.** The table below lists all possible outcomes and the corresponding x values and probabilities.

Outcome	x	Probability
FFFF	0	(0.8)(0.8)(0.8)(0.8) = 0.4096
FFFS	1	(0.8)(0.8)(0.8)(0.2) = 0.1024
FFSF	1	(0.8)(0.8)(0.2)(0.8) = 0.1024
FSFF	1	(0.8)(0.2)(0.8)(0.8) = 0.1024
SFFF	1	(0.2)(0.8)(0.8)(0.8) = 0.1024
FFSS	2	(0.8)(0.8)(0.2)(0.2) = 0.0256
FSFS	2	(0.8)(0.2)(0.8)(0.2) = 0.0256
FSSF	2	(0.8)(0.2)(0.2)(0.8) = 0.0256
SFFS	2	(0.2)(0.8)(0.8)(0.2) = 0.0256
SFSF	2	(0.2)(0.8)(0.2)(0.8) = 0.0256
SSFF	2	(0.2)(0.2)(0.8)(0.8) = 0.0256
FSSS	3	(0.8)(0.2)(0.2)(0.2) = 0.0064
SFSS	3	(0.2)(0.8)(0.2)(0.2) = 0.0064
SSFS	3	(0.2)(0.2)(0.8)(0.2) = 0.0064
SSSF	3	(0.2)(0.2)(0.2)(0.8) = 0.0064
SSSS	4	(0.2)(0.2)(0.2)(0.2) = 0.0016

From this we deduce the following probability distribution of x.

Value of x	Probability
0	0.4096
1	0.4096
2	0.1536
3	0.0256
4	0.0016

b. Both 0 and 1 are most likely values since each has probability 0.4096 of occurring.

c. P(at least 2 of the 4 have earthquake insurance) = P(x ≥ 2) = 0.1536 + 0.0256 + 0.0016 = 0.1808

7.17 **a.** The smallest value of y is 1. The outcome corresponding to this is S.
The second smallest value of y is 2. The outcome that corresponds to this is FS.

b. The set of all possible y values is the set of all positive integers.

c. P(y = 1) = P(S) = 0.7

P(y = 2) = P(FS) = (0.3)(0.7) = 0.21

P(y = 3) = P(FFS) = (0.3)(0.3)(0.7) = $(0.3)^2$ (0.7) = 0.063

P(y = 4) = P(FFFS) = (0.3)(0.3)(0.3)(0.7) = $(0.3)^3$ (0.7) = 0.0189

P(y = 5) = P(FFFFS) = (0.3)(0.3)(0.3)(0.3)(0.7) = $(0.3)^4$ (0.7) = 0.00567

We do see a pattern here. In fact, p(y) = P(y-1 failures followed by a success) = $(0.3)^{y-1}$ (0.7) for y = 1, 2, 3, ...

7.19 The following table lists all possible outcomes and the corresponding values of y.

Magazine 1 arrives on	Magazine 2 arrives on	Probability	Value of y
Wednesday	Wednesday	(0.4)(0.4) = 0.16	0
Wednesday	Thursday	(0.4)(0.3) = 0.12	1
Wednesday	Friday	(0.4)(0.2) = 0.08	2
Wednesday	Saturday	(0.4)(0.1) = 0.04	3
Thursday	Wednesday	(0.3)(0.4) = 0.12	1
Thursday	Thursday	(0.3)(0.3) = 0.09	1
Thursday	Friday	(0.3)(0.2) = 0.06	2
Thursday	Saturday	(0.3)(0.1) = 0.03	3
Friday	Wednesday	(0.2)(0.4) = 0.08	2
Friday	Thursday	(0.2)(0.3) = 0.06	2
Friday	Friday	(0.2)(0.2) = 0.04	2
Friday	Saturday	(0.2)(0.1) = 0.02	3
Saturday	Wednesday	(0.1)(0.4) = 0.04	3
Saturday	Thursday	(0.1)(0.3) = 0.03	3
Saturday	Friday	(0.1)(0.2) = 0.02	3
Saturday	Saturday	(0.1)(0.1) = 0.01	3

From the above table we deduce the following probability distribution for y.

Value of y	Probability
0	0.16
1	0.33
2	0.32
3	0.19

Exercises 7.21 – 7.25

7.21 **a.** P(at most 5 minutes elapses before dismissal) = (1/10)(5 – 0) = 0.5

  b. $P(3 \leq x \leq 5)$ = (1/10)(5 – 3) = 0.2

a. The density curve for x is shown below.

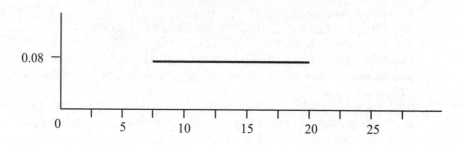

b. The area under the density curve between x = 7.5 and x = 20.0 must be 1. This implies that the height of the density curve = 1/(20 − 7.5) = 1/12.5 = 0.08.

c. P(x is at most 12) = (0.08)(12 − 7.5) = (0.08)(4.5) = 0.36.

d. P(x is between 10 and 15) = (0.08)(15 − 10) = 0.4.
P(x is between 12 and 17) = (0.08)(17 − 12) = 0.4.
The two probabilities are equal because the distribution is uniform and therefore the probabilities depend only on the length of the interval for which the probability is being sought. This ensures they have the same area.

7.25 **a.** The height of the density curve must be 1/20 = 0.05. So
P(x is less than 10 minutes) = (0.05)(10 − 0) = 0.5.
P(x is more than 15 minutes) = (0.05)(20 − 15) = 0.25.

b. P(x is between 7 and 12 minutes) = (0.05)(12 − 7) = 0.25.

c. P(x < c) = (0.05)(c) = 0.9 when c = 0.9/0.05 = 18. So the value of c is 18.

Exercises 7.27 − 7.43

7.27 **a.** $P(x \leq 0.5) = 0.5\left(\dfrac{0.5+1}{2}\right) = 0.5(0.75) = 0.375$

$P(0.25 \leq x \leq 0.5) = 0.25\left(\dfrac{0.75+1}{2}\right) = 0.25(0.875) = 0.21875$

$P(x \geq 0.75) = 0.25\left(\dfrac{1.25+1.5}{2}\right) = 0.25(1.375) = 0.34375$

b. The mean of x is $\dfrac{7}{12} = 0.583333$ and the standard deviation of x = $\sqrt{\dfrac{11}{144}} = 0.276385$. So,

P(x is more than one standard deviation from the mean value)
= P(x is less than (0.583333 − 0.276385)) + P(x is greater than (0.583333 + 0.276385))
= P(x is less than 0.306948) + P(x is greater than 0.859718)
= $(0.306948)\left(\dfrac{0.5+0.806948}{2}\right) + (1 - 0.859718)\left(\dfrac{1.359718+1.5}{2}\right) = 0.4012.$

7.29   **a.** $\mu_y = (0)(0.65) + (1)(0.20) + (2)(0.10) + (3)(0.04) + (4)(0.01) = 0.56$. In the long run, the average number of broken eggs per carton will equal 0.56.

   b. P(number of broken eggs is less than μ_y) = P(y is less than 0.56) = P(y = 0) = 0.65. So, in the long run, about 65% of the cartons will have fewer than μ_y broken eggs. This is not surprising because the distribution of y is skewed to the right and so the mean of y is greater than the median of y. We would expect the required proportion to be more than 0.5 because of the skewness.

   c. The indicated calculation would be correct if the values of y were all equally likely, but this is not the case. The values 0 and 1 occur more often than 2 or 3 or 4. Hence we need a "weighted average" rather than a simple average.

7.31 $\sigma_x^2 = (-1.2)^2 (0.54) + (-0.2)^2 (0.16) + (0.8)^2 (0.06) + (1.8)^2 (0.04) + (2.8)^2 (0.20) = 2.52$. Hence, $\sigma_x = \sqrt{2.52} = 1.5875$.

7.33   **a.** $\mu_x = (1)(0.05) + (2)(0.10) + (3)(0.12) + (4)(0.30) + (5)(0.30) + (6)(0.11) + (7)(0.01) + (8)(0.01) = 4.12$.

   b. Using the definition of variance in the textbook, we get $\sigma_x^2 = 1.94560$. So $\sigma_x = \sqrt{1.94560} = 1.3948$. The average squared distance of a value of x from its mean is 1.94560. The average distance of a value of x from its mean is approximately 1.3948.

   c. P(x is within 1 standard deviation of the mean)
   = P(x is between (4.12 – 1.3948) and (4.12 + 1.3948))
   = P(x is 3, 4 or 5) = 0.12 + 0.30 + 0.30 = 0.72.

   d. P(x is more than 2 standard deviations from the mean) = P(x is 1, 7, or 8)
   = 0.05 + 0.01 + 0.01 = 0.07.

7.35 $\mu_x = 10,550$. The author expects to get $ 10,550 under the royalty plan whereas the flat payment is for $10,000. It would appear that the author should choose the royalty plan if he/she were quite confident about his/her assessment of the probability distribution of x. On the other hand, P(x > 10,000) = 0.25 and P(x < 10000) = 0.35, so it is more likely that the royalty plan would yield an amount less than 10,000 dollars than an amount greater than 10,000 dollars, so if the author isn't sure about his/her assessment of the probability distribution of x, then he/she might prefer the flat payment plan.

7.37   **a.** y is a discrete random variable because there are only 6 possible values. Successive possible values have gaps between them (this is always the case when the variable takes on only a finite number of possible values).

   b. P(paid more than $1.20 per gallon) = 0.10 + 0.16 + 0.08 + 0.06 = 0.40.
   P(paid less than $1.40 per gallon) = 0.36 + 0.24 + 0.10 + 0.16 = 0.86.

   c. The mean value of y is 126.808 (cents per gallon) and the standard deviation is 13.3162 (cents per gallon). In the long run, the average value of y will be 126.808 cents/gallon and the deviation of y on any given day, from the mean value of y, will be about 13.3162 cents/gallon.

7.39 The variable y is related to the variable x of Problem **7.38** by the relation $y = 100 - 5x$. Hence the mean of y is $100 - 5(\text{Mean of } x) = 100 - (5)(2.3) = 88.5$. The variance of $y = (25)(\text{variance of } x) = (25)(0.81) = 20.25$.

7.41　**a.** Because if $y > 0, \Rightarrow x_2 > x_1 \Rightarrow$ the diameter of the peg > the diameter of the hole and the peg wouldn't fit in the hole!

b. $0.253 - 0.25 = 0.003$

c. $\sqrt{0.002^2 + 0.006^2} = 0.006$

d. Yes, it is reasonable to assume they are independent, they are made by different tools and are randomly selected.

e. With a standard deviation large than a mean, it seems fairly likely to obtain a negative value of y so it would seem a relatively common occurrence to find a peg that was too big to fit in the pre-drilled hole.

7.43　**a.** Mean of $x = 2.8$; standard deviation of $x = 1.289$.

b. Mean of $y = 0.7$; standard deviation of $y = 0.781$.

c. Let w_1 = total amount of money collected from cars. Then $w_1 = 3 x$.
Mean of $w_1 = 3$ (Mean of x) $= (3)(2.8) = 8.4$ dollars.
Variance of $w_1 = 9$ (Variance of x) $= (9)(1.66) = 14.94$.

d. Let w_2 = total amount of money collected from buses. Then $w_2 = 10 y$.
Mean of $w_2 = 10$ (Mean of y) $= (10)(0.7) = 7$ dollars.
Variance of $w_2 = 100$ (Variance of y) $= (100)(0.61) = 61$.

e. Let z = total number of vehicles on the ferry. Then $z = x + y$.
Mean of z = Mean of x + Mean of y $= 2.8 + 0.7 = 3.5$.
Variance of z = Variance of x + Variance of y $= 1.66 + 0.61 = 2.27$.

f. $w = w_1 + w_2$, so Mean of w = Mean of w_1 + Mean of $w_2 = 8.4 + 7 = 15.4$ dollars.
Variance of w = Variance of w_1 + Variance of $w_2 = 14.94 + 61 = 75.94$.

Exercises 7.45 – 7.63

7.45　**a.** There are exactly 6 such outcomes. They are SFFFFF, FSFFFF, FFSFFF, FFFSFF, FFFFSF, FFFFFS.

b. In a binomial experiment consisting of 20 trials, the number of outcomes with exactly 10 S's is equal to $\binom{20}{10} = 184756$. The number of outcomes with exactly 15 S's is equal to

$\binom{20}{15} = 15504$. The number of outcomes with exactly 5 S's is also equal to 15504 because

$\binom{20}{15} = \binom{20}{5}$.

7.47　**a.** $p(4) = \binom{6}{4}\pi^4(1-\pi)^{6-4} = (15)(0.8)^4(0.2)^2 = 0.24576$. This means, in the long run, in samples of 6 passengers selected from passengers flying a long route, the proportion of the time exactly 4 out of the 6 will sleep or rest will be close to 0.24576.

b. $p(6) = (0.8)^6 = 0.262144$.

c. $p(x \geq 4) = p(x = 4) + p(x = 5) + p(x = 6) = 0.245760 + 0.393216 + 0.262144 = 0.90112$.

7.49 **a.** $p(2) = \binom{5}{2} \pi^2 (1 - \pi)^{5-2} = (10)(0.25)^2 (0.75)^3 = 0.26367$.

b. $P(x \leq 1) = p(0) + p(1) = 0.2373046875 + 0.3955078125 = 0.6328125$.

c. $P(2 \leq x) = 1 - P(x \leq 1) = 1 - 0.6328125 = 0.3671875$.

d. $P(x \neq 2) = 1 - P(x = 2) = 1 - p(2) = 1 - 0.26367 = 0.73633$.

7.51 **a.** $P(X = 10) = (0.85)^{10} = 0.1969$

b. $P(X \leq 8) = 0.4557$

c. $p = 0.15$, $n = 500$ mean = 75, st. dev. = 7.984

d. 25 is more than 3 standard deviations from the mean value of x, so yes, this is a surprising result.

7.53 Suppose the graphologist is just guessing, i.e., deciding which handwriting is which by simply a coin toss. Then there is a 50% chance of guessing correctly in a single test. The probability of getting 6 or more correct in 10 trials, simply by guessing = $p(6) + p(7) + p(8) + p(9) + p(10)$, where $p(x)$ is the probability that a binomial random variable with $n = 10$ and $\pi = 0.5$ will take the value x. Using Appendix Table 9, we find this probability to be = $0.205 + 0.117 + 0.044 + 0.010 + 0.001 = 0.377$. Therefore, correctly guessing 6 or more out of 10 is not all that rare even without any special abilities. So, the evidence given here is certainly not convincing enough to conclude that the graphologist has any special abilities.

7.55 **a.** P(at most 5 fail inspection) = $p(0) + p(1) + p(2) + p(3) + p(4) + p(5)$, where $p(x)$ is the probability that a binomial random variable with $n = 15$ and $\pi = 0.3$ will take the value x. Using Appendix Table 9, we get P(at most 5 fail inspection) = $0.005 + 0.030 + 0.092 + 0.170 + 0.218 + 0.207 = 0.722$.

b. P(between 5 and 10 (inclusive) fail inspection) = $p(5) + p(6) + p(7) + p(8) + p(9) + p(10)$ = $0.207 + 0.147 + 0.081 + 0.035 + 0.011 + 0.003 = 0.484$.

c. Here, let x = number of cars that pass inspection. Then x is a binomial random variable with $n = 25$ and $\pi = 1 - 0.3 = 0.7$. Hence the expected value of x is $(25)(0.7) = 17.5$ and the standard deviation is $\sqrt{25(0.7)(1 - 0.7)} = 2.2913$.

7.57 Here n/N = 1000/10000 = 0.1 which is greater than 0.05. So a binomial distribution is not a good approximation for x = number of invalid signatures in a sample of size 1000 since the sampling is done without replacement.

7.59 **a.** P(program is implemented | $\pi = 0.8$) = $P(x \leq 15)$ where x is a binomial random variable with $n = 25$ and $\pi = 0.8$. Using Appendix Table 9, we calculate this probability to be 0.17.

b. P(program not implemented | $\pi = 0.7$) = $P(x > 15)$ where x is a binomial random variable with $n = 25$ and $\pi = 0.7$. Using Appendix Table 9, we calculate this probability to be 0.811.

P(program not implemented | $\pi = 0.6$) = $P(x > 15)$ where x is a binomial random variable with $n = 25$ and $\pi = 0.6$. Using Appendix Table 9, we calculate this probability to be 0.425.

c. Suppose the value 15 is changed to 14 in the decision criterion.

Then P(program is implemented | $\pi = 0.8$) = P(x $\leq$ 14) where x is a binomial random variable with n = 25 and $\pi = 0.8$. Using Appendix Table 9, we calculate this probability to be 0.0.006.

P(program not implemented | $\pi = 0.7$) = P(x > 14) where x is a binomial random variable with n = 25 and $\pi = 0.7$. Using Appendix Table 9, we calculate this probability to be 0.902.

P(program not implemented | $\pi = 0.6$) = P(x > 14) where x is a binomial random variable with n = 25 and $\pi = 0.6$. Using Appendix Table 9, we calculate this probability to be 0.586. The modified decision criterion leads to a lower probability of implementing the program when it need not be implemented and a higher probability of not implementing the program when it should be implemented.

7.61   **a.** Geometric distribution. We are not counting the number of successes in a fixed number of trials; instead, we are counting the number of trials needed to achieve a single success.

b. P(exactly two tosses) = (0.9)(0.1) = 0.09.

c. P(more than three tosses will be required) = P(first three attempts are failures) = $(0.9)^3$ = 0.729.

7.63   **a.** Geometric distribution

b. P(X = 3) = 0.1084

c. P(X < 4) = P(X $\leq$ 3) = 0.3859

d. P(X > 3) = 1 – P(X $\leq$ 3) = 1 - 0.3859 = 0.6141

Exercises 7.64 – 7.80

7.65   **a.** (z < –1.28) = 0.1003

b. (z > 1.28) = 1 – P(z $\leq$ 1.28) = 1 – 0.8997 = 0.1003

c. (–1 < z < 2) = P(z < 2) – P(z < –1) = 0.9772 – 0.1587 = 0.8185

d. (z > 0) = 1 – P(z $\leq$ 0) = 1 – 0.5 = 0.5

e. (z > –5) = 1 – P(z $\leq$ –5) $\approx$ 1 – 0 = 1

f. (–1.6 < z < 2.5) = P(z < 2.5) – P(z < –1.6) = 0.9938 – 0.0548 = 0.9390

g. (z < 0.23) = 0.5910

7.67   **a.** (z < 0.1) = 0.5398

b. (z < –0.1) = 0.4602

c. (0.40 < z < 0.85) = P(z < 0.85) – P(z < 0.4) = 0.8023 – 0.6554 = 0.1469

d. P(–0.85 < z < –0.40) = P(z < –0.4) – P(z < –0.85) = 0.3446 – 0.1977 = 0.1469

e. P(–0.40 < z < 0.85) = P(z < 0.85) – P(z < –0.4) = 0.8023 – 0.3446 = 0.4577

f. P(–1.25 < z) = 1 – P(z $\leq$ –1.25) = 1 – 0.1056 = 0.8944

g. P(z < –1.5 or z > 2.5) = P(z < –1.5) + 1 – P(z $\leq$ 2.5) = 0.0668 + 1 – 0.9938 = 0.0730

7.69   **a.** P(z > z*) = 0.03 $\Rightarrow$ z* = 1.88

b. P(z > z*) = 0.01 $\Rightarrow$ z* = 2.33

c. P(z < z*) = 0.04 $\Rightarrow$ z* = –1.75

d. P(z < z*) = 0.10 $\Rightarrow$ z* = –1.28

7.71 **a.** 91st percentile = 1.34
b. 77th percentile = 0.74
c. 50th percentile = 0
d. 9th percentile = −1.34
e. They are negatives of one another. The 100pth and 100(1−p)th percentiles will be negatives of one another, because the z curve is symmetric about 0.

7.73 **a.** $P(x > 4000) = P(z > \dfrac{4000 - 3432}{482}) = P(z > 1.1784) = 0.1193$

$P(3000 \le x \le 4000) = P(\dfrac{3000 - 3432}{482} \le z \le \dfrac{4000 - 3432}{482})$

$= P(-0.8963 \le z \le 1.1784) = P(z \le 1.1784) - P(z < -0.8963) = 0.8807 - 0.1851 = 0.6956$

b. $P(x < 2000) + P(x > 5000) = P(z < \dfrac{2000 - 3432}{482}) + P(z > \dfrac{5000 - 3432}{482})$

$= P(z < -2.97095) + P(z > 3.25311) = 0.0015 + 0.00057 = 0.00207$

c. $P(x > 7 \text{ lbs}) = P(x > 7(453.6) \text{ grams}) = P(x > 3175.2) = P(z > \dfrac{3175.2 - 3432}{482})$

$= P(z > -0.53278) = 0.70191$

d. We find x_1^* and x_2^* such that $P(x < x_1^*) = 0.0005$ and $P(x > x_2^*) = 0.0005$. The most extreme 0.1% of all birthweights would then be characterized as weights less than x_1^* or weights greater than x_2^*. $P(x < x_1^*) = P(z < z_1^*) = 0.0005$ implies that $z_1^* = -3.2905$. So $x_1^* = \mu + z_1^* \sigma = 3432 + 482(-3.2905) = 1846$ grams. $P(x > x_1^*) = P(z > z_2^*) = 0.0005$ implies that $z_2^* = 3.2905$. So $x_2^* = \mu + z_2^* \sigma = 3432 + 482(3.2905) = 5018$ grams. Hence the most extreme 0.1% of all birthweights correspond to weights less than 1846 grams or weights greater than 5018 grams.

e. If x is a random variable with a normal distribution and a is a numerical constant (not equal to 0) then $y = ax$ also has a normal distribution. Furthermore, mean of y = $a\times$ (mean of x) and standard deviation of y = $a\times$ (standard deviation of x). The birth weight distribution would be normal with mean 7.5663 lbs and standard deviation 1.06263 lbs. The probability that the birth weight is greater than 7 lbs. is still 0.7019.
Let y be the birthweights measured in pounds. Recalling that one pound = 453.6 grams, we have $y = \left(\dfrac{1}{453.6}\right) x$, so $a = \dfrac{1}{453.6}$. The distribution of y is normal with mean equal to 3432/453.6 = 7.56614 pounds and standard deviation equal to 482/453.6 = 1.06261 pounds. So $P(y > 7 \text{ lbs}) = P(z > \dfrac{7 - 7.56614}{1.06261}) = P(z > -0.53278) = 0.70291$. As expected, this is the same answer that we obtained in part **c.**

7.75 For the second machine, P(that a cork doesn't meet specifications) = 1 − P(2.9 ≤ x ≤ 3.1)

$= 1 - P(\dfrac{2.9 - 3.05}{0.01} \le z \le \dfrac{3.1 - 3.05}{0.01}) = 1 - P(-15 \le z \le 5) \approx 1 - 1 = 0.$

7.77 $\dfrac{c-120}{120} = -1.28 \Rightarrow c = 120 - 1.28(20) = 94.4$

Task times of 94.4 seconds or less qualify an individual for the training.

7.79 **a.** $P(x \le 60) = P(z \le \dfrac{60-60}{15}) = P(z \le 0) = 0.5$

$P(x < 60) = P(z < 0) = 0.5$

b. $P(45 < x < 90) = P(\dfrac{45-60}{15} < z < \dfrac{90-60}{15})$

$= P(-1 < z < 2) = 0.9772 - 0.1587 = 0.8185$

c. $P(x \ge 105) = P(z \ge \dfrac{105-60}{15}) = P(z \ge 3) = 1 - P(z < 3) = 1 - 0.9987 = 0.0013$

The probability of a typist in this population having a net rate in excess of 105 is only 0.0013. Hence it would be surprising if a randomly selected typist from this population had a net rate in excess of 105.

d. $P(x > 75) = P(z > \dfrac{75-60}{15}) = P(z > 1) = 1 - P(z \le 1) = 1 - 0.8413 = 0.1587$

$P(\text{both exceed } 75) = (0.1587)(0.1587) = 0.0252$

e. $P(z < z^*) = 0.20 \Rightarrow z^* = -0.84$

$x^* = \mu + z^*\sigma \Rightarrow x^* = 60 + (-0.84)(15) = 60 - 12.6 = 47.4$

So typing speeds of 47.4 words or less per minute would qualify individuals for this training.

Exercises 7.81 – 7.91

7.81 Since this plot appears to be very much like a straight line, it is reasonable to conclude that the normal distribution provides an adequate description of the steam rate distribution.

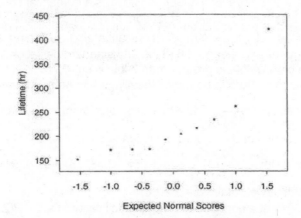

7.83

Since the graph exhibits a pattern substantially different from that of a straight line, one would conclude that the distribution of the variable "component lifetime" cannot be adequately modeled by a normal distribution. It is worthwhile noting that this "deviation from normality" could be due to the single outlying value of 422.6.

7.85

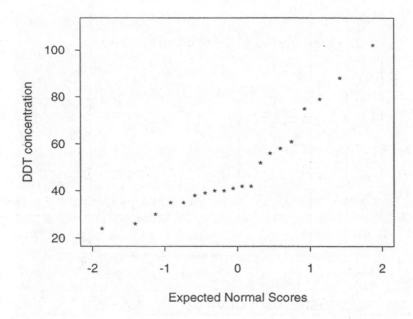

Although the graph follows a straight line pattern approximately, there is a distinct "kink" in the graph at about the value 45 on the vertical axis. Points corresponding to DDT concentration less than 45 seem to follow one straight line pattern while those to the right of 45 seem to follow a different straight line pattern. A normal distribution may not be an appropriate model for this population.

7.87 Histograms of the square root transformed data as well as the cube root transformed data are given below. It appears that the histogram of the cube root transformed data is more symmetric than the histogram of the square root transformed data. (However, keep in mind that the shapes of these histograms are somewhat dependent on the choice of class intervals.)

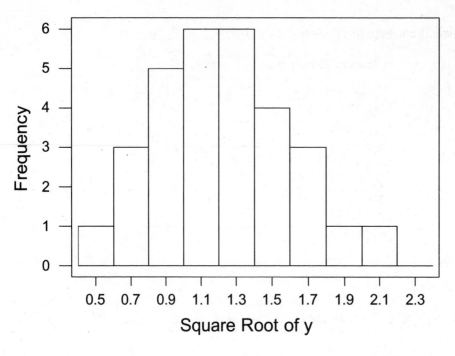

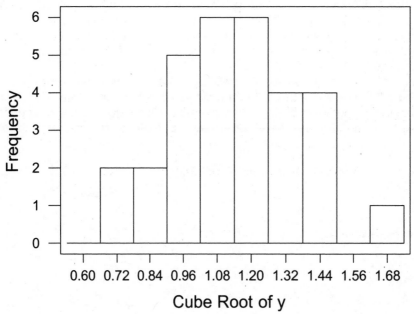

7.89   **a.** The required frequency distribution is given below.

class	frequency	relative frequency
0-<100	22	0.22
100-<200	32	0.32
200-<300	26	0.26
300-<400	11	0.11
400-<500	4	0.04
500-<600	3	0.03
600-<700	1	0.01
700-<800	0	0
800-<900	1	0.01

b.

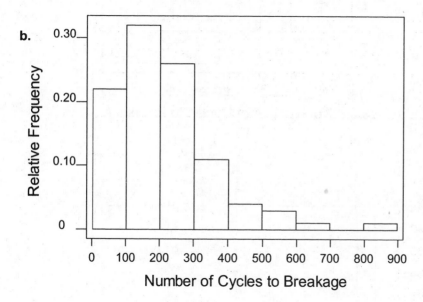

This histogram is positively skewed.

c. Data were square root transformed and the corresponding relative frequency histogram is shown below. Clearly the distribution of the transformed data is more symmetric than that of the original data.

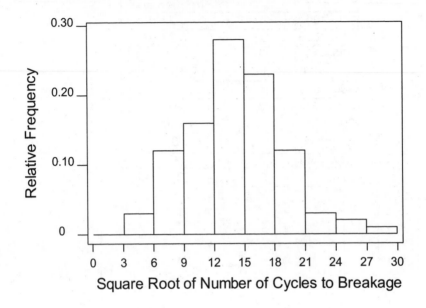

Square Root of Number of Cycles to Breakage

7.91   a.

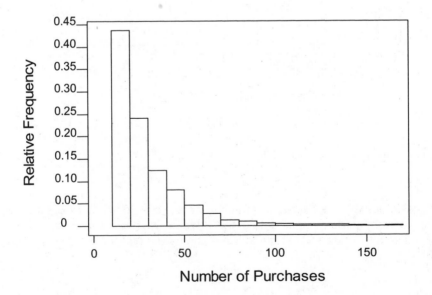

Number of Purchases

0	56667889999
1	000011122456679999
2	0011111368
3	11246
4	18
5	
6	8
7	
8	
9	39 HI: 448

b. A density histogram of the body mass data is shown below. It is positively skewed. The scale on the horizontal axis extends upto body mass = 500 in order to accommodate the HI value of 448.

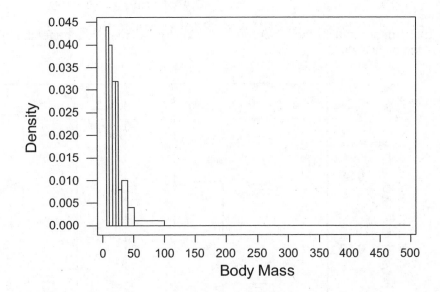

If statistical inference procedures based on normality assumptions are to be used to draw conclusions from the body mass data of this problem, then a transformation should be considered so that, on the transformed scale, the data are approximately normally distributed.

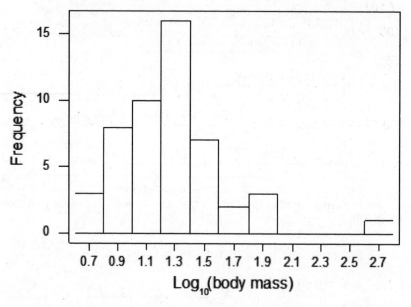

The normal curve fits the histogram of the log transformed data better than the histogram of the original data.

d.

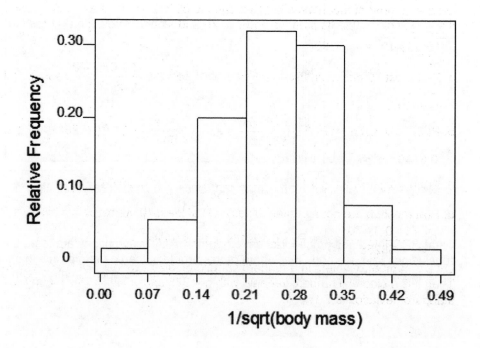

The histogram in this transformed scale does appear to have an approximate bell-shape, i.e., in the transformed scale, the data do appear to follow a normal distribution model.

Exercises 7.93 – 7.101

7.93 **a.** $P(x = 100) = P\left(\dfrac{100-0.5-100}{15} \le z \le \dfrac{100+0.5-100}{15}\right) = P(-0.03333 \le z \le 0.03333)$

= 0.02659.

b. $P(x \le 110) = P\left(z \le \dfrac{110+0.5-100}{15}\right) = P(z \le 0.7) = .75804.$

c. $P(x < 110) = P(x \le 109) = P\left(z \le \dfrac{109+0.5-100}{15}\right) = P(z \le 0.6333) = 0.73674.$

d. $P(75 \le x \le 125) = P\left(\dfrac{75-0.5-100}{15} \le z \le \dfrac{125+0.5-100}{15}\right) = P(-1.7 \le z \le 1.7) =$

0.9109.

7.95 **a.** $P(650 \le x) = P\left(\dfrac{650-0.5-500}{75} \le z\right) = P(1.9933 \le z) = 0.02311.$

b. $P(400 < x < 550) = P(401 \le x \le 549) = P\left(\dfrac{401-0.5-500}{75} \le z \le \dfrac{549+0.5-500}{75}\right)$

= P(-1.3267 ≤ z ≤ 0.66) = 0.6531.

c. $P(400 \le x \le 550) = P\left(\dfrac{400-0.5-500}{75} \le z \le \dfrac{550+0.5-500}{75}\right)$

= P(-1.34 ≤ z ≤ 0.6733) = 0.6595.

7.97 Let x = number of mountain bikes in a sample of 100. Then x has a binomial distribution with n = 100 and $\pi = 0.7$. Its mean is equal to 70 and standard deviation is equal to

$\sqrt{100(0.7)(0.3)}$ = 4.58258.

a. $P(\text{at most 75 mountain bikes}) = P(x \le 75) = P\left(z \le \dfrac{75+0.5-70}{4.58258}\right) = P(z \le 1.2002) =$

0.88497.

b. $P(60 \le x \le 75) = P\left(\dfrac{60-0.5-70}{4.58258} \le z \le \dfrac{75+0.5-70}{4.58258}\right) = P(-2.29129 \le z \le 1.2002) =$

0.8740.

c. $P(80 < x) = P(81 \le x) = P\left(\dfrac{81-0.5-70}{4.58258} \le z\right) = P(2.29129 \le z) = 0.01097.$

d. P(at most 30 are not mountain bikes) = P(at least 70 are mountain bikes) = $P(70 \le x) =$

$P\left(\dfrac{70-0.5-70}{4.58258} \le z\right) = P(-0.1091 \le z) = 0.54344.$

7.99 Let x = number of voters in a random sample of size 225 that favors a 7-day waiting period. Then x is a binomial random variable with n = 225 and $\pi = 0.65$. Its mean is 146.25 and standard deviation is 7.15454.

a. $P(150 \leq x) = P\left(\dfrac{150 - 0.5 - 146.25}{7.15454} \leq z\right) = P(0.45426 \leq z) = 0.32482.$

b. $P(150 < x) = P(151 \leq x) = P\left(\dfrac{151 - 0.5 - 146.25}{7.15454} \leq z\right) = P(.59403 \leq z) = 0.27625.$

c. $P(x < 125) = P(x \leq 124) = \left(z \leq \dfrac{124 + 0.5 - 146.25}{7.15454}\right) = P(z \leq -3.04003) = 0.001183.$

7.101   **a.** Let x = number of mufflers replaced under the warranty in a random sample of 400 purchases. Then x is binomial with n = 400 and $\pi = 0.2$ Its mean is 80 and standard deviation is 8.

$P(75 \leq x \leq 100) = P\left(\dfrac{75 - 0.5 - 80}{8} \leq z \leq \dfrac{100 + 0.5 - 80}{8}\right) = P(-0.6875 \leq z \leq 2.5625) =$

0.74892.

b. $P(x \leq 70) = P\left(z \leq \dfrac{70 + 0.5 - 80}{8}\right) = P(z \leq -1.1875) = 0.11752.$

c. $P(x < 50) = P(x \leq 49) = P\left(z \leq \dfrac{49 + 0.5 - 80}{8}\right) = P(z \leq -3.8125) = 0.00006878.$ This

probability is so close to zero that, if indeed the 20% figure were true, it is highly unlikely that fewer than 50 mufflers among the 400 randomly selected purchases were ever replaced. Therefore the 20% figure is highly suspect.

Exercises 7.103 – 7.123

7.103   Let x = number of customers out of the 15 who want the diet Coke. Then x is binomial with n = 15 and $\pi = 0.6$. P(each of the 15 is able to purchase the drink desired) = P($5 \leq x \leq 10$) = 0.025 + 0.061 + 0.118 + 0.177 + 0.207 + 0.196 = 0.784 (using Appendix Table 9).

7.105   **a.** Mean of x = 2.64000; standard deviation = 1.53961.

b. P(x is farther than 3 standard deviations from its mean)
= P(x is less than -1.97883 OR x is greater than 7.25883) = 0.

7.107   **a.** $P(0.525 \leq y \leq 0.550) = P\left(\dfrac{0.525 - 0.5}{0.025} \leq z \leq \dfrac{0.575 - 0.5}{0.025}\right) = P(1 \leq z \leq 2) = 0.1359.$

b. P(y exceeds its mean value by more than 2 standard deviations) = P($2 < z$) = 0.02275.

c. Let π = P(a randomly selected pizza has at least 0.475 lb of cheese) = P($0.475 \leq y$)
P($-1 \leq z$) = 0.84134. Let x = number of pizzas in a random sample of size 3 that have at least 0.475 lb of cheese. Then x is a binomial random variable with n = 3 and π = 0.84134. Therefore the probability that all 3 chosen pizzas have at least 0.475 lb of cheese = P(x = 3) = $(0.84134)^3$ = 0.59555.

7.109   **a.** P(y > 45) = P(z > –1.5) = 0.9332.

b. Let K denote the amount exceeded by only 10% of all clients at a first meeting.

Then $0.10 = P(y > K) = P\left(z > \dfrac{K - 60}{10}\right)$ implies that $\dfrac{K - 60}{10} = 1.2816$. So K = 72.816

minutes.

c. Let R = revenue. Then $R = 10 + 50\left(\dfrac{y}{60}\right)$. So mean value of $R = 10 + 50\left(\dfrac{\text{Mean of } y}{60}\right) =$

$10 + 50\left(\dfrac{60}{60}\right) = 60$ dollars.

7.111 $P(x < 4.9) = P(z < \dfrac{4.9 - 5}{0.05}) = P(z < -2) = 0.0228.$ $P(5.2 < x) = P(\dfrac{5.2 - 5}{0.05} < z) = P(4 < z) \approx 0$

7.113 a. $P(x < 5'7") = P(x < 67") = P(z < \dfrac{67 - 66}{2}) = P(z < 0.5) = 0.6915$

No, the claim that 94% of all women are shorter than 5'7" is not correct. Only about 69% of all women are shorter than 5'7".

b. About 69% of adult women would be excluded from employment due to the height requirement.

7.115 a. The 36 possible outcomes and the corresponding values of w are listed in the following table. The notation (i, j) means that Allison arrived at time i P.M and Teri arrived at time j P.M.

Outcome	w	Outcome	w	Outcome	w	Outcome	w
(1,1)	0	(2,4)	2	(4,1)	3	(5,4)	1
(1,2)	1	(2,5)	3	(4,2)	2	(5,5)	0
(1,3)	2	(2,6)	4	(4,3)	1	(5,6)	1
(1,4)	3	(3,1)	2	(4,4)	0	(6,1)	5
(1,5)	4	(3,2)	1	(4,5)	1	(6,2)	4
(1,6)	5	(3,3)	0	(4,6)	2	(6,3)	3
(2,1)	1	(3,4)	1	(5,1)	4	(6,4)	2
(2,2)	0	(3,5)	2	(5,2)	3	(6,5)	1
(2,3)	1	(3,6)	3	(5,3)	2	(6,6)	0

From the above table we deduce the following probability distribution of w.

w	p(w)
0	6/36
1	10/36
2	8/36
3	6/36
4	4/36
5	2/36

b. The expected value of w is equal to 70/36 = 1.9444 hours.

7.117 a. We use L to denote that Lygia won a game and B to denote Bob won a game. By a sequence such as BLLB we mean that Bob won the first and the fourth games and Lygia won the second and the third games. Then x = 4 occurs if either BBBB or LLLL occurs. So $p(4) = (0.6)^4 + (0.4)^4 = 0.1552$.

b. The outcomes that lead to x = 5 are, LLLBL, LLBLL, LBLLL, BLLLL, BBBLB, BBLBB, BLBBB, LBBBB. There for $p(5) = 4 (0.6)^4(0.4) + 4 (0.6)(0.4)^4 = 0.2688$.

103

c. The maximum value of x is 7 and the minimum value of x is 4. The following table gives the distribution of x.

x	p(x)
4	$(0.6)^4 + (0.4)^4 = 0.1552$
5	$\binom{4}{3}(0.6)^4(0.4) + \binom{4}{3}(0.6)(0.4)^4 = 0.26880$
6	$\binom{5}{3}(0.6)^4(0.4)^2 + \binom{5}{3}(0.6)^2(0.4)^4 = 0.29952$
7	$\binom{6}{3}(0.6)^4(0.4)^3 + \binom{6}{3}(0.6)^3(0.4)^4 = 0.27648$

d. The expected value of x is equal to $(4)(0.1552)+(5)(0.2688) + (6)(0.29952) + (7)(0.27648)$
= 5.69728.

7.119 x = number among three randomly selected customers who buy brand W. In the following, we use the sequence such as DWP to mean that the first customer buys brand D, the second buys brand W, and the third buys brand P, etc.
$P(x = 0) = $ P(each of the three customers failed to buy brand W) $= (1-0.4)^3 = 0.216$.
$P(x = 1) = $ P(exactly one of 3 customers bought brand W) $= 3(0.4)(0.6)^2 = 0.432$.
$P(x = 2) = $ P(exactly two of 3 customers bought brand W) $= 3(0.4)^2(0.6) = 0.288$.
$P(x = 3) = $ P(all three bought brand W) $= (0.4)^3 = 0.064$.

7.121 a. $P(5.9 < x < 6.15) = P(\dfrac{5.9-6}{0.1} < z < \dfrac{6.15-6}{0.1}) = P(-1 < z < 1.5)$
= $0.9332 - 0.1587 = 0.7745$

b. $P(6.1 < x) = P(\dfrac{6.1-6}{0.1} < z) = P(1 < z) = 1 - P(z \le 1) = 1 - 0.8413 = 0.1587$

c. $P(x < 5.95) = P(z < \dfrac{5.95-6}{0.1}) = P(z < -0.5) = 0.3085$

d. The largest 5% of the pH values are those pH values which exceed the 95th percentile. The 95th percentile is $6 + 1.645(0.1) = 6.1645$.

7.123 a. Let x= number among the 200 who are uninsured. Then x is binomial with n= 200 and $\pi = 0.16$. Its mean is $(200)(0.16) = 32$ and standard deviation is $\sqrt{(200)(0.16)(1-0.16)} = 5.18459$.

b. $P(25 \le x \le 40) = P\left(\dfrac{25-0.5-32}{5.18459} \le z \le \dfrac{40+0.5-32}{5.18459}\right) = P(-1.44659 \le z \le 1.63947) = 0.87544.$

c. $P(x > 50) = P(51 \le x) = P\left(\dfrac{51-0.5-32}{5.18459} \le z\right) = P(3.56827 \le z) = 0.0001797$. This probability is very close to zero, so if this outcome occurs, one would be led to doubt the 16% figure.

Chapter 8

Exercises 8.1 – 8.13

8.1   A statistic is any quantity computed from the observations in a sample. A population characteristic is a quantity which describes the population from which the sample was taken.

8.3   **a.** population characteristic
b. statistic
c. population characteristic
d. statistic
e. statistic

8.5   Ten additional random samples of size 5 and their sample means are:

Sample	Sample mean
323, 261, 159, 222, 342	261.4
168, 184, 323, 230, 261	233.2
159, 295, 323, 261, 275	262.6
262, 231, 263, 222, 168	229.2
275, 261, 270, 159, 319	256.8
159, 261, 323, 258, 267	253.6
262, 168, 261, 333, 275	259.8
267, 222, 295, 231, 261	255.2
168, 230, 267, 184, 262	222.2
231, 323, 159, 261, 265	247.8

The value of $\bar{x}$ differs from one sample to the next. Samples 1 and 3 produced $\bar{x}$ values greater than   while the remaining eight samples yielded $\bar{x}$ values less than   . Samples 2, 4, and 9 produced $\bar{x}$ values which differ from   by quite a large amount, while samples 1, 3, 5, 6, 7, 8, and 10 produced $\bar{x}$ values quite close to   . These $\bar{x}$ values are in general agreement with the results summarized in Figure 8.4 of the text. One area of disagreement may be that 7 of the ten values of $\bar{x}$ were less than   , whereas in Figure 8.4, 22 out of 45 (roughly half) of the $\bar{x}$ values were less than   .

8.7   **a.**

Sample	Sample Mean	Sample	Sample Mean
1, 2	1.5	3, 1	2.0
1, 3	2.0	3, 2	2.5
1, 4	2.5	3, 4	3.5
2, 1	1.5	4, 1	2.5
2, 3	2.5	4, 2	3.0
2, 4	3.0	4, 3	3.5

The sampling distribution of $\bar{x}$ based on n = 2, when sampling is done <u>without replacement</u>, is

Value of $\bar{x}$	1.5	2.0	2.5	3.0	3.5
Probability	2/12	2/12	4/12	2/12	2/12

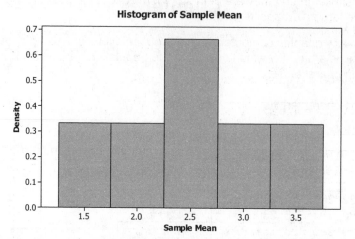

Histogram of Sample Mean

b.

Sample	Sample Mean	Sample	Sample Mean
1, 1	1.0	3, 1	2.0
1, 2	1.5	3, 2	2.5
1, 3	2.0	3, 3	3.0
1, 4	2.5	3, 4	3.5
2, 1	1.5	4, 1	2.5
2, 2	2.0	4, 2	3.0
2, 3	2.5	4, 3	3.5
2, 4	3.0	4, 4	4.0

The sampling distribution of $\bar{x}$ based on n = 2, when sampling is done with replacement, is

Value of $\bar{x}$	1.0	1.5	2.0	2.5	3.0	3.5	4.0
Probability	1/16	2/16	3/16	4/16	3/16	2/16	1/16

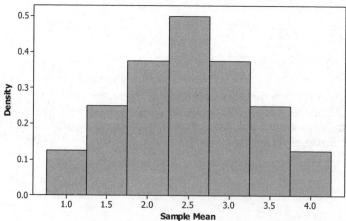

Histogram of Sample Mean

c. They are similar in that they are both symmetric about the population mean of 2.5. Also, both have largest probability at 2.5.

They are different in that, the values in **a** cover a smaller range, 1.5 to 3.5, than in part **b**, 1.0 to 4.0. Also, the probability that $\overline{x}$ will take a value within 0.5 unit of the true mean is 8/12 in part **a** whereas it is only 10/16 (a smaller value) in part **b**.

8.9   **Note:** Statistic #3 is sometimes called the **midrange**.

Sample	Value of Mean	Value of Median	Value of statistic #3 (midrange)
2, 3, 3*	2.67	3	2.5
2, 3, 4	3.00	3	3.0
2, 3, 4*	3.00	3	3.0
2, 3*, 4	3.00	3	3.0
2, 3*, 4*	3.00	3	3.0
2, 4, 4*	3.33	4	3.0
3, 3*, 4	3.33	3	3.5
3, 3*, 4*	3.33	3	3.5
3, 4, 4*	3.67	4	3.5
3*, 4, 4*	3.67	4	3.5

Sampling distribution of statistic #1.

Value of $\overline{x}$	2.67	3	3.33	3.67
Probability	0.1	0.4	0.3	0.2

Sampling distribution of statistic #2.

Value of median	3	4
Probability	0.7	0.3

Sampling distribution of statistic #3.

Value of midrange	2.5	3	3.5
Probability	0.1	0.5	0.4

Statistic #1 is the only one that is unbiased, but because it has a large standard deviation, there is a good chance that an estimate would be far away from the value of the population mean. For this population, Statistic #3 may be a better choice because it is more likely to produce an estimate that is close to the population mean.

Exercises 8.11 – 8.21

8.11 For n = 36, 50, 100, and 400

8.13 a. $\mu_{\bar{x}} = 40, \sigma_{\bar{x}} = \dfrac{5}{\sqrt{64}} = \dfrac{5}{8} = 0.625$

Since n = 64, which exceeds 30, the shape of the sampling distribution will be approximately normal.

b. $P(\mu - 0.5 < \bar{x} < \mu + 0.5) = P(39.5 < \bar{x} < 40.5) = P(\dfrac{39.5 - 40}{0.625} < z < \dfrac{40.5 - 40}{0.625})$

$= P(-0.8 < z < 0.8) = 0.7881 - 0.2119 = 0.5762$

c. $P(\bar{x} < 39.3 \text{ or } \bar{x} > 40.7) = P(z < \dfrac{39.3 - 40}{0.625} \text{ or } z > \dfrac{40.7 - 40}{0.625})$

$= 1 - P(-1.12 < z < 1.12) = 1 - [0.8686 - 0.1314] = 0.2628$

8.15 a. $\mu_{\bar{x}} = 2$ and $\sigma_{\bar{x}} = \dfrac{0.8}{\sqrt{9}} = 0.267$

b. For n = 20, $\mu_{\bar{x}} = 2$ and $\sigma_{\bar{x}} = \dfrac{0.8}{\sqrt{20}} = 0.179$

For n = 100, $\mu_{\bar{x}} = 2$ and $\sigma_{\bar{x}} = \dfrac{0.8}{\sqrt{100}} = 0.08$

In all three cases $\bar{x}$ has the same value, but the standard deviation of $\bar{x}$ decreases as n increases. A sample of size 100 would be most likely to result in an $\bar{x}$ value close to . This is because the sampling distribution of $\bar{x}$ for n = 100 has less variability than those for n = 9 or 20.

8.17 a. $\sigma_{\bar{x}} = \dfrac{5}{\sqrt{25}} = 1,$

$P(64 \le \bar{x} \le 67) = P(\dfrac{64 - 65}{1} \le z \le \dfrac{67 - 65}{1}) = P(-1 \le z \le 2) = 0.9772 - 0.1587 = 0.8185$

$P(68 \le \bar{x}) = P(\dfrac{68 - 65}{1} \le z) = P(3 \le z) = 1 - P(z < 3) = 1 - 0.9987 = 0.0013$

b. Because the sample size is large we can use the normal approximation for the sampling distribution of $\bar{x}$ by the central limit theorem. We have

$\sigma_{\bar{x}} = \dfrac{5}{\sqrt{100}} = 0.5,$

$P(64 \le \bar{x} \le 67) = P((64 - 65)/0.5 < z < (67 - 65)/0.5) = P(-2 < z < 4)$
$= 1 - 0.0228 = 0.9772$
$P(68 \le \bar{x}) = P((68 - 65)/.5 \le z) = P(6 \le z) = 1 - P(z < 6) \approx 1 - 1 = 0$

8.19 If the true process mean is equal to 0.5 in, then $\mu_{\bar{x}} = 0.5, \sigma_{\bar{x}} = \dfrac{0.02}{\sqrt{36}} = 0.00333$

P(the line will be shut down unnecessarily) = $1 - P(0.49 \le \bar{x} \le 0.51)$

$= 1 - P\left(\dfrac{0.49 - 0.50}{0.00333} \le z \le \dfrac{0.51 - 0.50}{0.00333}\right)$

$= 1 - P(-3 \le z \le 3) = 1 - (0.9987 - 0.0013) = 0.0026$

8.21 The total weight of their baggage will exceed the limit, if the average weight exceeds $6000/100 = 60$.

$\mu_{\bar{x}} = 50, \sigma_{\bar{x}} = \dfrac{20}{\sqrt{100}} = 2$,

P(total weight $\ge$ 6000) = $P(\bar{x} \ge 60) = P\left(z \ge \dfrac{60-50}{2}\right) = P(z \ge 5) = 1 - P(z < 5) \approx 1 - 1 = 0$

Exercises 8.23 – 8.31

8.23   **a.** $\mu_p = 0.65, \sigma_p = \sqrt{0.65(0.35)/10} = 0.15083$

    b. $\mu_p = 0.65, \sigma_p = \sqrt{0.65(0.35)/20} = 0.10665$

    c. $\mu_p = 0.65, \sigma_p = \sqrt{0.65(0.35)/30} = 0.08708$

    d. $\mu_p = 0.65, \sigma_p = \sqrt{0.65(0.35)/50} = 0.06745$

    e. $\mu_p = 0.65, \sigma_p = \sqrt{0.65(0.35)/100} = 0.04770$

    f. $\mu_p = 0.65, \sigma_p = \sqrt{0.65(0.35)/200} = 0.03373$

8.25   **a.** $\mu_p = 0.07, \sigma_p = \sqrt{\dfrac{(0.07)(0.93)}{100}} = 0.0255$

    b. No; $np = 100(0.07) = 7$, and $n(1-p) = 100(1 - 0.07) = 93$. For the sampling distribution of p to be considered approximately normal, both have to be greater or equal to 10.

    c. The value of the mean doesn't change as it isn't dependent on the sample size. The standard deviation becomes : $\sigma_p = \sqrt{\dfrac{(0.07)(0.93)}{200}} = 0.01804$

    d. Yes; $np = 200(0.07) = 14$, and $n(1-p) = 200(1 - 0.07) = 186$. For the sampling distribution of p to be considered approximately normal, both have to be greater or equal to 10.

    e. $P(p > 0.1) = P\left(z > \dfrac{0.1 - 0.07}{0.01804}\right) = P(z > 1.66) = 0.0485$

8.27   **a.** $\mu_p = 0.005, \sigma_p = \sqrt{\dfrac{(0.005)(0.995)}{100}} = 0.007$

    b. Since $np = 100(0.005) = 0.5$ is less than 10, the sampling distribution of p cannot be approximated well by a normal curve.

    c. The requirement is that $np \ge 10$, which means that n would have to be at least 2000.

8.29 **a.** For $\pi = 0.5$, $\mu_p = 0.5$ and $\sigma_p = \sqrt{\dfrac{0.5(0.5)}{225}} = 0.0333$

For $\pi = 0.6$, $\mu_p = 0.6$ and $\sigma_p = \sqrt{\dfrac{0.6(0.4)}{225}} = 0.0327$

For both cases, $n\pi \geq 10$ and $n(1-\pi) \geq 10$. Hence, in each instance, p would have an approximately normal distribution.

b. For $\pi = 0.5$, $P(p \geq 0.6) = P(z \geq \dfrac{0.6 - 0.5}{0.0333}) = P(z \geq 3) = 1 - P(z < 3)$

$= 1 - 0.9987 = 0.0013$.

For $\pi = 0.6$, $P(p \geq 0.6) = P(z \geq \dfrac{0.6 - 0.6}{0.0327}) = P(z \geq 0) = 1 - P(z < 0)$

$= 1 - 0.5000 = 0.5000$

c. When $\pi = 0.5$, the $P(p \geq 0.6)$ would decrease.
When $\pi = 0.6$, the $P(p \geq 0.6)$ would remain the same.

8.31 **a.** $\mu_p = \pi = 0.05$, $\sigma_p = \sqrt{\dfrac{(0.05)(0.95)}{200}} = 0.01541$

$P(p > 0.02) = P(z > \dfrac{0.02 - 0.05}{0.01541}) = P(z > -1.95)$

$= 1 - P(z \leq -1.95) \approx 1 - 0.0258 = 0.9742$

b. $\mu_p = \pi = 0.10$, $\sigma_p = \sqrt{\dfrac{(0.1)(0.9)}{200}} = 0.02121$

$P(p \leq 0.02) = P(z \leq \dfrac{0.02 - 0.10}{0.02121}) = P(z \leq -3.77) \approx 0$

Exercises 8.33 – 8.37

8.33 **a.** The sampling distribution of $\bar{x}$ will be approximately normal, with mean equal to 50 (lb)

and standard deviation equal to $\dfrac{1}{\sqrt{100}} = 0.1$ (lb).

b. $P(49.75 < \bar{x} < 50.25) = P(\dfrac{49.75 - 50}{0.1} < z < \dfrac{50.25 - 50}{0.1}) = P(-2.5 < z < 2.5)$

$= .9938 - .0062 = .9876$

c. $P(\bar{x} < 50) = P(z < \dfrac{50 - 50}{0.1}) = P(z < 0) = 0.5$

8.35 **a.** $P(850 < x < 1300) = P(\dfrac{850 - 1000}{150} < z < \dfrac{1300 - 1000}{150}) = P(-1 < z < 2)$

$= 0.9772 - 0.1587 = 0.8185$

b. $\mu_{\bar{x}} = 1000, \sigma_{\bar{x}} = \dfrac{150}{\sqrt{10}} = 47.43$

$P(950 < \bar{x} < 1100) = P(\dfrac{950 - 1000}{47.43} < z < \dfrac{1100 - 1000}{47.43})$

$= P(\ 1.054 < z < 2.108) = 0.9825 \quad 0.1459 = 0.8366$

$P(850 < \bar{x} < 1300) = P(\dfrac{850 - 1000}{47.43} < z < \dfrac{1300 - 1000}{47.43})$

$= P(\ 3.16 < z < 6.33) = 1 \quad 0.0008 = 0.9992$

8.37 $\mu = 100, \sigma = 30, \mu_{\bar{x}} = 100, \sigma_{\bar{x}} = \dfrac{30}{\sqrt{50}} = 4.2426$

$P(\text{ total} > 5300) = P(\bar{x} > 106) = P(z > \dfrac{106 - 100}{4.2426}) = P(z > 1.4142 < z) = 1 \quad P(z \quad 1.4142)$

$= 1 \quad 0.9214 = 0.0786$

Chapter 9

Exercises 9.1 – 9.9

9.1 Statistic II would be preferred because it is unbiased and has smaller variance than the other two.

9.3 $p = \dfrac{1720}{6212} = 0.2769$

9.5 The point estimate of   would be $p = \dfrac{\text{number in sample registered}}{n} = \dfrac{14}{20} = 0.70$

9.7 a. $\dfrac{19.57}{10} = 1.957$

  b. $s^2 = 0.15945$

  c. $s = 0.3993$; No, this estimate is not unbiased. It underestimates the true value of   .

9.9 a. The value of   will be estimated by using the statistic s. For this sample,

   $\Sigma x^2 = 1757.54, \Sigma x = 143.6, n = 12$

$$s^2 = \dfrac{\Sigma x^2 - \dfrac{(\Sigma x)^2}{n}}{n-1} = \dfrac{1757.54 - \dfrac{(143.6)^2}{12}}{12-1} = \dfrac{1757.54 - 1718.4133}{11}$$

$$= \dfrac{39.1267}{11} = 3.557 \text{ and } s = \sqrt{3.557} = 1.886$$

  b. The population median will be estimated by the sample median. Since $n = 12$ is even, the sample median equals the average of the middle two values (6th and 7th values), i.e.,

   $\dfrac{(11.3 + 11.4)}{2} = 11.35$.

  c. In this instance, a trimmed mean will be used. First arrange the data in increasing order. Then, trimming one observation from each end will yield an 8.3% trimmed mean. The trimmed mean equals 117.3/10 = 11.73.

  d. The point estimate of   would be $\bar{x} = 11.967$. From part a, $s = 1.886$. Therefore the estimate of the 90th percentile is 11.967 + 1.28(1.886) = 14.381.

Exercises 9.11 – 9.29

9.11 a. As the confidence level increases, the width of the large sample confidence interval also increases.

  b. As the sample size increases, the width of the large sample confidence interval decreases.

9.13 For the interval to be appropriate, np 10, $n(1$ $p)$ 10 must be satisfied.

  a. $np = 50(0.3) = 15$, $n(1$ $p) = 50(0.7) = 35$, yes

  b. $np = 50(0.05) = 2.5$, no

  c. $np = 15(0.45) = 6.75$, no

  d. $np = 100(0.01) = 1$, no

  e. $np = 100(0.70) = 70$, $n(1$ $p) = 100(0.3) = 30$, yes

  f. $np = 40(0.25) = 10$, $n(1$ $p) = 40(0.75) = 30$, yes

g. $np = 60(0.25) = 15$, $n(1-p) = 60(0.75) = 45$, yes

h. $np = 80(0.10) = 8$, no

9.15 **a.** Because $np = 420.42$ and $n(1-p) = 580.58$, which are both greater than 10, and the Americans in the sample were randomly selected from a large population, the large-sample interval can be used. The sample proportion p is 0.42. The 99% confidence interval for π, the proportion of all Americans who made plans in May 2005 based on an incorrect weather report would be $0.42 \pm 2.58\sqrt{\dfrac{0.42(1-0.42)}{1001}}$ $\quad 0.42 \quad 0.0.0402 \quad$ (0.3798, 0.4602). We can be 99% confident that the true proportion of adults Americans who made plans in May 2005 based on an incorrect weather report is between .38 and .46.

b. No, weather reports may be more or less reliable during other months.

9.17 **a.** Because np and $n(1-p)$ are both greater than 10, and the potential jurors in the sample were randomly selected from a large population, the large-sample interval can be used. The sample proportion p is 350/500 = 0.7. The 95% confidence interval for π, the population proportion would be $0.7 \pm 1.96\sqrt{\dfrac{0.7(1-0.7)}{500}}$ $\quad 0.7 \quad 0.0402 \quad$ (0.6598, 0.7402). With 95% confidence we can estimate that between 66% and 74% of all potential jurors regularly watch at least one crime-scene investigation series.

b. A 99% confidence interval would be wider than the 95% confidence interval in Part (a).

9.19 **a.** Because np and $n(1-p)$ are both greater than 10, and the businesses in the sample were randomly selected from a large population, the large-sample interval can be used. The sample proportion p is 137/526 = 0.26. The 95% confidence interval would be $0.26 \pm 1.96\sqrt{\dfrac{0.26(1-0.26)}{526}}$ $\quad 0.26 \quad 0.0375 \quad$ (0.2225, 0.2975). With 95% confidence we can estimate that between 22.25% and 29.75% of all U.S. businesses have fired workers for misuse of the Internet.

b. It would be narrower because of a lower confidence level (90% instead of 95%) and because of a smaller estimated standard error (0.0189 instead of 0.0191).

9.21 Bound of error (based on 95% confidence) = $1.96\sqrt{\dfrac{p(1-p)}{n}}$. If $p = 0.82$ and $n = 1002$, $1.96\sqrt{\dfrac{0.82(1-0.82)}{1002}} = 0.0238$. This implies a 95% confidence interval of 0.82 ± 0.0238 or (0.7724, 0.8676). With 95% confidence, we can estimate that between 77.2% and 86.8% of all adults think that reality shows are 'totally made up' or 'mostly distorted'.

9.23 **a.** The sample proportion p is 38/115 = 0.330. The 95% confidence interval would be $0.330 \pm 1.96\sqrt{\dfrac{0.330(1-0.330)}{115}}$ $\quad 0.330 \quad 0.0859 \quad$ (0.2441, 0.4159).

b. The sample proportion p is 22/115 = 0.1913. The 90% confidence interval would be $0.1913 \pm 1.645\sqrt{\dfrac{0.1913(1-0.1913)}{115}}$ $\quad 0.1913 \quad 0.0603 \quad$ (0.1310, 0.2516).

114

c. The interval is wider in part a because (i) the confidence interval is higher in part (a) and (ii) the sample proportion is more extreme; further from 0.5 in part (a) .

9.25 The sample proportion p is 18/52 = 0.3462. The 95% confidence interval would be

$$0.3462 \pm 1.96\sqrt{\frac{0.3462(1-0.3462)}{52}} \qquad 0.3462 \quad 0.1293 \quad (.2169, .4755).$$

The assumption are: p is the sample proportion from a random sample, and that the sample size is large, (np 10, and $n(1-p)$ 10).

9.27 With $p = .25$ and $n = 1002$, the bound on error is $1.96\sqrt{\frac{(.25)(.75)}{1002}} \approx .03$

9.29 $n = 0.25\left[\frac{1.96}{B}\right]^2 = 0.25\left[\frac{1.96}{0.05}\right]^2 = 384.16$;   take $n = 385$.

Exercises 9.31 – 9.49

9.31 **a.** 2.12
b. 1.80
c. 2.81
d. 1.71
e. 1.78
f. 2.26

9.33 As the sample size increases, the width of the interval decreases. The interval (51.3, 52.7) has a width of 52.7 51.3 = 1.4 and the interval (49.4, 50.6) has a width of 50.6 49.4 = 1.2. Hence, the interval (49.4, 50.6) is based on the larger sample size.

9.35 Because the specimens were randomly selected and the distribution of the breaking force is approximately normal, the t confidence interval formula for the mean can be used.   The 95% confidence interval for μ , the population mean breaking force is:

$$306.09 \pm 1.96\left(\frac{41.97}{\sqrt{6}}\right) \Rightarrow 306.09 \pm 33.58 \Rightarrow (272.51, 339.67).$$ With 95% confidence we can estimate

the average breaking force for acrylic bone cement to be between 272.5 and 229.7 Newtons.

9.37 Because the adults were randomly selected and the sample size is large, the t confidence interval formula for the mean can be used.   The 90% confidence interval for μ , the

population mean commute time is: $28.5 \pm 1.645\frac{24.2}{\sqrt{500}} \Rightarrow 28.5 \pm 1.780 \Rightarrow (26.72, 30.28)$.

With 90% confidence, we estimate that the mean daily commute time for all working residents of Calgary, Canada is between 26.7 minutes and 30.3 minutes.

9.39 **a.** The t critical value for a 95% confidence interval when df = 99 is 1.99. The confidence interval based on this sample data is

$$\bar{x} \pm (t\ critical)\frac{s}{\sqrt{n}} \Rightarrow 183 \pm (1.99)\left(\frac{20}{\sqrt{100}}\right) \Rightarrow (179.03, 186.97).$$

b. $\bar{x} \pm (t\ critical)\frac{s}{\sqrt{n}} \Rightarrow 190 \pm (1.99)\left(\frac{23}{\sqrt{100}}\right) \Rightarrow (185.44, 194.56)$ i

c. The new FAA recommendations are above the upper level of both confidence levels so it appears that Frontier airlines have nothing to worry about.

9.41 **a.** The t critical value for a 90% confidence interval when df = 9 is 1.83. The confidence interval based on this sample data is

$$\bar{x} \pm (t \text{ critical})\frac{s}{\sqrt{n}} \Rightarrow 54.2 \pm (1.83)\left(\frac{3.6757}{\sqrt{10}}\right) \Rightarrow 54.2 \pm 2.1271 \Rightarrow (52.073, 56.327).$$

b. If the same sampling method was used to obtain other samples of the same size and confidence intervals were calculated from these samples, 90% of them would contain the true population mean.

c. As airlines are often rated by how often their flights are late, I would recommend the published arrival time to be close to the upper bound of the confidence interval of the journey time: 10: 57 a.m.

9.43 A boxplot (below) shows the distribution to be slightly skewed with no outliers. It seems plausible that the population distribution is approximately normal. Calculation of a confidence interval for the population mean cadence requires sample mean and sample standard deviation:

$$\bar{x} = 0.926 \qquad s = 0.0809 \quad \text{t critical value with 19 df is 2.58.}$$

The interval is $0.926 \pm (2.86)\left(\frac{0.0809}{\sqrt{20}}\right) = 0.926 \pm 0.052 = (0.874, 0.978)$. With 99%

confidence, we estimate the mean cadence of all healthy men to be between 0.874 and 0.978 strides per second.

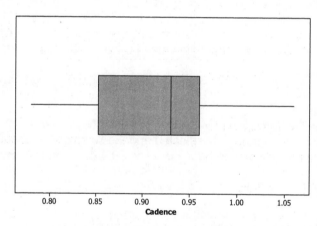

9.45   Summary statistics for the sample are: n   5, x̄   17, s   9.03

The 95% confidence interval is given by

$$\bar{x} \pm (t\ critical)\frac{s}{\sqrt{n}} \Rightarrow 17 \pm (2.78)\frac{9.03}{\sqrt{5}} \Rightarrow 17 \pm 11.23 \Rightarrow (5.77, 28.23).$$

9.47   Since the sample size is small ($n = 17$), it would be reasonable to use the t confidence interval only if the population distribution is normal (at least approximately). A histogram of the sample data (see figure below) suggests that the normality assumption is not reasonable for these data. In particular, the values 270 and 290 are much larger than the rest of the data and the distribution is skewed to the right. Under the circumstances the use of the t confidence interval for this problem is not reasonable.

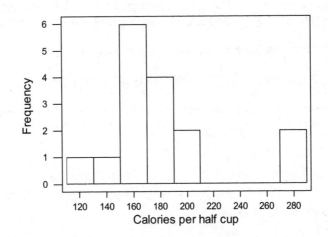

9.49 $n = \left[\dfrac{(z\ critical)\sigma}{B}\right]^2 = \left[\dfrac{(1.96)(1)}{0.1}\right]^2 = (19.6)^2 = 384.16$. Hence, n should be 385.

Exercises 9.51 – 9.73

9.51 $p = \dfrac{466}{1014} = 0.4596$

9.53   A 90% confidence interval is $0.65 \pm 1.645\sqrt{\dfrac{0.65(1-0.65)}{150}} \Rightarrow 0.65 \pm 0.064 \Rightarrow (0.589, 0.714)$

Thus, we can be 90% confident that between 58.9% and 71.4% of Utah residents favor fluoridation. This is consistent with the statement that a clear majority of Utah residents favor fluoridation.

9.55 **a.** $25.62 \pm 2.33\left(\dfrac{14.41}{\sqrt{44}}\right) \Rightarrow 25.62 \pm 5.062 \Rightarrow (20.558, 30.682)$

b. $18.10 \pm 2.33\left(\dfrac{15.31}{\sqrt{257}}\right) \Rightarrow 18.10 \pm 2.225 \Rightarrow (15.875, 20.325)$

c. It is based on a larger sample.

d. Since the interval in **a** gives the plausible values for   and the lower endpoint is greater than 20, this suggests that the mean number of hours worked per week for non-persistors is greater than 20.

9.57 Using a conservative value of $\pi = .5$ in the formula for required sample size gives:

$n = \pi(1-\pi)\left(\dfrac{1.96}{B}\right)^2 = .25\left(\dfrac{1.96}{.05}\right)^2 = 384.16$. A sample size of at least 385 should be used.

9.59 $n = 0.25\left[\dfrac{2.576}{B}\right]^2 = 0.25\left[\dfrac{2.576}{0.10}\right]^2 = 165.9$; take $n = 166$.

9.61 **a.** The 95% confidence interval for $_{ABC}$ is

$15.6 \pm (1.96)\left(\dfrac{5}{\sqrt{50}}\right) \Rightarrow 15.6 \pm 1.39 \Rightarrow (14.21, 16.99)$.

b. For μ_{CBS} : $11.9 \pm 1.39 \Rightarrow (10.51, 13.29)$

For μ_{FOX} : $11.7 \pm 1.39 \Rightarrow (10.31, 13.09)$

For μ_{NBC} : $11.0 \pm 1.39 \Rightarrow (9.61, 12.39)$

c. Yes, because the plausible values for $_{ABC}$ are larger than the plausible values for the other means. That is, $_{ABC}$ is plausibly at least 14.21, while the other means are plausibly no greater than 13.29, 13.09, and 12.39.

9.63 Since $n = 18$, the degrees of freedom is $n - 1 = 18 - 1 = 17$.

From the t-table, the t critical value is 2.11.

The confidence interval is $9.7 \pm 2.11\left(\dfrac{4.3}{\sqrt{18}}\right) \Rightarrow 9.7 \pm 2.14 \Rightarrow (7.56, 11.84)$.

Based on this interval we can conclude with 95% confidence that the true average rating of acceptable load is between 7.56 kg and 11.84 kg.

9.65 $B = 0.1$, $= 0.8$

$n = \left[\dfrac{1.96\sigma}{B}\right]^2 = \left[\dfrac{(1.96)(0.8)}{0.1}\right]^2 = (15.68)^2 = 245.86$

Since a partial observation cannot be taken, n should be *rounded up* to $n = 246$.

9.67 The 99% confidence interval for the mean commuting distance based on this sample is

$$\bar{x} \pm (t \text{ critical})\frac{s}{\sqrt{n}} \Rightarrow 10.9 \pm (2.58)\left(\frac{6.2}{\sqrt{300}}\right) \Rightarrow 10.9 \pm 0.924 \Rightarrow (9.976, 11.824).$$

9.69 Example 9.8 has shown that the sample data meets the conditions for the *t* confidence interval.

99% upper confidence bound for the true mean wait time for bypass patients in Ontario:

$$\bar{x} + 2.33 \frac{s}{\sqrt{n}} = 19 + 2.33\left(\frac{10}{\sqrt{539}}\right) = 19 + 1.004 = 20.004$$

9.71 The 95% confidence interval for the population standard deviation of commuting distance is

$$6.2 \pm 1.96\left(\frac{6.2}{\sqrt{2(300)}}\right) \Rightarrow 6.2 \pm 0.496 \Rightarrow (5.704, 6.696).$$

9.73 The t critical value for a 90% confidence interval when df = 12   1 = 11 is 1.80. The confidence interval based on this sample data is

$$\bar{x} \pm (t \text{ critical})\frac{s}{\sqrt{n}} \Rightarrow 21.9 \pm (1.80)\left(\frac{7.7}{\sqrt{12}}\right) \Rightarrow 21.9 \pm 4.00 \Rightarrow (17.9, 25.9).$$

With 90% confidence, the mean time to consume a frog by Indian False Vampire bats is between 17.9 and 25.9 minutes.

Chapter 10

Exercises 10.1 – 10.11

10.1 $\bar{x} = 50$ is not a legitimate hypothesis, because $\bar{x}$ is a statistic, not a population characteristic. Hypotheses are always expressed in terms of a population characteristic, not in terms of a sample statistic.

10.3 If we use the hypothesis H_o: $\mu = 100$ versus H_a: $\mu > 100$, we are taking the position that the welds do not meet specifications, and hence, are not acceptable unless there is substantial evidence to show that the welds are good (i.e. $\mu > 100$). If we use the hypothesis H_o: $\mu = 100$ versus H_a: $\mu < 100$, we initially believe the welds to be acceptable, and hence, they will be declared unacceptable only if there is substantial evidence to show that the welds are faulty. It seems clear that we would choose the first set-up, which places the burden of proof on the welding contractor to show that the welds meet specifications.

10.5 Not being able to conclude that the MMR does cause autism is not the same as concluding that the MMR does not cause autism – just as failing to reject the null hypothesis is not the same as accepting the null hypothesis.

10.7 H_o: $p = 0.5$ versus H_a: $p > 0.5$

10.9 A majority is defined to be more than 50%. Therefore, the commissioner should test:
$$H_o: p = 0.5 \quad \text{versus} \quad H_a: p > 0.5$$

10.11 Since the manufacturer is interested in detecting values of μ which are less than 40, as well as values of μ which exceed 40, the appropriate hypotheses are: H_o: $\mu = 40$ versus H_a: $\mu \neq 40$

Exercises 10.13 – 10.21

10.13 a. Type I: Concluding the symptoms are due to a disease when, in truth, the symptoms are due to child abuse.
Type II: Concluding the symptoms are due to child abuse when, in truth, the symptoms are due to a disease.
b. Based on the quote, the doctor considers a Type I error to be more serious.

10.15 a. Pizza Hut's decision is consistent with the decision of rejecting H_o.
b. Rejecting H_o when it is true is called a type I error. So if they incorrectly reject H_o, they are making a type I error.

10.17 a. A type I error is returning to the supplier a shipment which is not of inferior quality.
A type II error is accepting a shipment of inferior quality.
b. The calculator manufacturer would most likely consider a type II error more serious, since they would then end up producing defective calculators.
c. From the supplier's point of view, a type I error would be more serious, because the supplier would end up having lost the profits from the sale of the good printed circuits.

10.19 a. The manufacturer claims that the percentage of defective flares is 10%. Certainly one would not object if the proportion of defective flares is less than 10%. Thus, one's primary concern would be if the proportion of defective flares exceeds the value stated by the manufacturer.

b. A type I error entails concluding that the proportion of defective flares exceeds 10% when, in fact, the proportion is 10% or less. The consequence of this decision would be the filing of charges of false advertising against the manufacturer, who is not guilty of such actions. A type II error entails concluding that the proportion of defective flares is 10% when, in reality, the proportion is in excess of 10%. The consequence of this decision is to allow the manufacturer who is guilty of false advertising to continue bilking the consumer.

10.21 **a.** They failed to reject the null hypothesis, because their conclusion "There is no evidence of increased risk of death due to cancer" for those living in areas with nuclear facilities is precisely what the null hypothesis states.

b. They would be making a type II error since a type I error is failing to reject the null hypothesis when the null is false.

c. Since the null hypothesis is the initially favored hypothesis and is presumed to be the case until it is determined to be false, if we fail to reject the null hypothesis, it is not proven to be true. There is just not sufficient evidence to refute its presumed truth. On the other hand, if the hypothesis test is conducted with

H_o :   is greater than the value for areas without nuclear facilities

H_a :   is less than or equal to the value for areas without nuclear facilities

and the null hypothesis is rejected, then there would be evidence based on data against the presumption of an increased cancer risk associated with living near a nuclear power plant. This is as close as one can come to "proving" the absence of an increased risk using statistical studies.

Exercises 10.23 – 10.43

10.23   The P-value is the probability, assuming H_o, of obtaining a test statistic value at least as contradictory to H_o as what actually resulted.

a. A P-value of 0.0003 indicates that these test results are very unlikely to occur if H_o is true. The more plausible explanation for the occurrence of these results is that H_o is false.

b. A P-value of 0.35 indicates that these results are quite likely to occur (consistent with) if H_o is true. The data do not cast reasonable doubt on the validity of H_o.

10.25 H_o is rejected if P-value   . H_o should be rejected for only the following pair:

   d: P-value = 0.084,   = 0.10

10.27   **a.** n = 25(0.2) = 5. n   10, The large sample test is not appropriate.

b. n = 10(0.6) = 6. n   10,   The large sample test is not appropriate.

c. n = 100(0.9) = 90, n(1   ) = 100(0.1) = 10, Both   10. The large sample test is appropriate.

d. n = 75(0.05) = 3.25. n   10,   The large sample test is not appropriate.

10.29   Let   represent the proportion of adults Americans who would favor the drafting of women.

   H_o: = 0.5

   H_a: < 0.5

We will compute a P-value for this test.   = 0.05.

Since the population is much larger than the sample, n = 1000(0.5)   10, and n(1   ) = 1000(0.5)   10, the given sample is a random sample and the sample size is large, the large sample z test may be used.

n = 1000, p = .43

$$z = \frac{0.43 - 0.5}{\sqrt{\dfrac{0.5(0.5)}{1000}}} = -4.43$$

P-value = 2(area under the z curve to the left of -4.43) ≈ 0

Since the P-value is less than   , H_0 is rejected.

There is enough evidence to suggest that less than half of all adult Americans would favor the drafting of women.

10.31 Let   represent the proportion of adult Americans who plan to alter their shopping habit if gas prices remain high.

  H_o: = 0.75
  H_a: > 0.75

Even though the problem doesn't state what value is to be used for   , for illustrative purposes we will use   = 0.05

Since the population is much larger than the sample, n   = 1813(0.75)   10, and n(1   ) = 1813(0.25)   10, the given sample is a random sample and the sample size is large, the large sample z test may be used.

n = 1813, p = 1432/1813 = 0.79

$$z = \frac{0.79 - .75}{\sqrt{\dfrac{0.75(0.25)}{1813}}} = 3.93$$

P-value = area under the z curve to the right of 3.93 ≈ 0

Since the P-value is less than   of 0.01, H_0 is rejected.

There is enough evidence to suggest that more than three-quarters of adult Americans plan to alter their shopping habits if gas prices remain high.

10.33 Let   represent the proportion of U.S. adults who believe that playing the lottery is the best strategy for accumulating $200,000 in net wealth.

  H_o: = 0.2
  H_a: > 0.2

Even though the problem doesn't state what value is to be used for   , for illustrative purposes we will use   = 0.05

Test Statistic: $z = \dfrac{p - \text{hypothesized value}}{\sqrt{\dfrac{(\text{hypothesized value})(1 - \text{hypothesized value})}{n}}} = \dfrac{p - .2}{\sqrt{\dfrac{(.2)(.8)}{n}}}$

Assumptions: This test requires a random sample and a large sample size. The given sample was a random sample, the population size is much larger than the sample size, and the sample size was n = 1000. Because 1000 (.2) ≥ 10 and 1000(.8) ≥ 10, the large-sample test is appropriate.

n = 1000, p = .21

$$z = \frac{0.21 - 0.2}{\sqrt{\frac{0.2(0.8)}{1000}}} = 0.79$$

P-value = area under the z curve to the right of 0.79 = 0.2148

Since the P-value is greater than   , H_0 is not rejected.

The data does not provide enough evidence to suggest that more than 20% of all adult Americans believe that playing the lottery is the best strategy for accumulating $200,000 in net wealth.

10.35 a. Let   represent the proportion of adult Americans who think that the quality of movies being produced is getting worse.

H_o: = 0.5

H_a: < 0.5

  = 0.05

Test Statistic: $z = \dfrac{p - \text{hypothesized value}}{\sqrt{\dfrac{(\text{hypothesized value})(1 - \text{hypothesized value})}{n}}} = \dfrac{p - .5}{\sqrt{\dfrac{(.5)(.5)}{n}}}$

Assumptions: This test requires a random sample and a large sample size. The given sample was a random sample, the population size is much larger than the sample size, and the sample size was n = 1000. Because 1000 (.5) ≥ 10 and 1000(1-.5) ≥ 10, the large-sample test is appropriate.

n = 1000, p = .47

$$z = \frac{0.47 - 0.5}{\sqrt{\frac{0.5(0.5)}{1000}}} = -1.90$$

P-value = area under the z curve to the left of -1.90 = .0287

Since the P-value is less than   , H_o is rejected. The data provides enough evidence to suggest that fewer than half of American adults believe that movie quality is getting worse.

b. The hypothesis test would be the identical up to the assumptions. The sample size is 100. Because 100(.5) ≥ 10 and 100(1-.5) ≥ 10, the large-sample test is appropriate.

n = 100, p = .47

$$z = \frac{0.47 - 0.5}{\sqrt{\frac{0.5(0.5)}{100}}} = -0.6$$

P-value = area under the z curve to the left of -0.6 = .2743

Since the P-value is greater than   , H_0 is not rejected.

The data does not provide enough evidence to suggest that fewer than half of American adults believe that movie quality is getting worse.

c. The denominator, σ_p: $\sqrt{\dfrac{\text{(hypothesized value)}(1 - \text{hypothesized value})}{n}}$ has the same

hypothesized value for both but for part (a), n = 1000, and for part (b), n = 100. The value of the test statistic for part (b) is larger which leads to a larger P-value. The P-value in (a) is smaller than α and the P-value in (b) is larger than α which leads to different conclusions. With a small sample the difference between the sample proportion of 0.47 and the hypothesized proportion of 0.5 could plausibly be attributed to chance, whereas when the sample size is larger this difference is no longer likely to be attributable to chance.

10.37 Let   represent the proportion of all baseball fans that think that the designated hitter rule should either be expanded to both baseball leagues or eliminated.

H_o: = 0.5
H_a: > 0.5

Even though the problem doesn't state what value is to be used for   , for illustrative purposes we will use   = 0.05

Test Statistic: $z = \dfrac{p - \text{hypothesized value}}{\sqrt{\dfrac{\text{(hypothesized value)}(1 - \text{hypothesized value})}{n}}} = \dfrac{p - .5}{\sqrt{\dfrac{(.5)(.5)}{n}}}$

Assumptions: This test requires a random sample and a large sample size. The given sample was a random sample, the population size is much larger than the sample size, and the sample size was n =394. Because 394(.5) ≥ 10 and 394(1 - .5) ≥ 10, the large-sample test is appropriate.

n = 394, p = 272/394 = .69

$z = \dfrac{0.69 - 0.5}{\sqrt{\dfrac{0.5(0.5)}{394}}} = 7.54$

P-value = area under the z curve to the right of 7.54   0

Since the P-value is less than   , H_0 is rejected.

The data provides enough evidence to suggest that the majority of all baseball fans think that the designated hitter rule should either be expanded to both baseball leagues or eliminated.

10.39 Let   represent the proportion of U.S. adults who believe that rudeness is a worsening problem.

H_o: = 0.75
H_a: > 0.75

We will compute a P-value for this test.   = 0.05.

Since n   = 2013(0.75)   10, and n(1   ) = 2013(0.25)   10, the large sample z test may be used.

n = 2013, p = 1283/2013 = 0.637357

$$z = \frac{0.63736 - 0.75}{\sqrt{\dfrac{0.75(0.25)}{2013}}} = \frac{-.11264}{.00965} = -11.67$$

P-value = area under the z curve to the right of –11.67 $\approx$ 1

Since the P-value is greater than α, H_o cannot be rejected.

There is not enough evidence to suggest that over ¾ of U.S. adults think that rudeness is a worsening problem.

10.41 Let π represent the proportion of religion surfers who belong to a religious community

$\quad\quad\quad H_o$: $\pi = 0.68$

$\quad\quad\quad H_a$: $\pi \neq 0.68$

We will compute a P-value for this test. $\alpha = 0.05$.

Since $n\pi = 512(0.68) \geq 10$, and $n(1-\pi) = 512(0.32) \geq 10$, the large sample z test may be used.

n = 512, p = .84

$$z = \frac{0.84 - .68}{\sqrt{\dfrac{0.68(0.32)}{512}}} = \frac{.16}{.0206} = 7.77$$

P-value = 2(area under the z curve to the right of 7.77) $\approx$ 2(0) = 0

Since the P-value is less than α, H_o is rejected.

There is enough evidence to suggest that the proportion of religion surfers that belong to a religious community is different than 68%.

10.43 Let π represent the proportion of the time the Belgium Euro lands with its head side up.

$\quad\quad\quad H_o$: $\pi = 0.5$

$\quad\quad\quad H_a$: $\pi \neq 0.5$

We will compute a P-value for this test. $\alpha = 0.01$.

Since $n\pi = 250(0.5) \geq 10$, and $n(1-\pi) = 250(0.5) \geq 10$, the large sample z test may be used.

n = 250, p = 140/250 = 0.56

$$z = \frac{0.56 - .5}{\sqrt{\dfrac{0.5(0.5)}{250}}} = \frac{.06}{.0316} = 1.90$$

P-value = 2(area under the z curve to the right of 1.90) $\approx$ 2(0.0287) = 0.0574

Since the P-value is greater than α of 0.01, H_o is not rejected.

There is not enough evidence to suggest that the proportion of the time that the Belgium Euro coin would land with its head side up is not 0.5. With a significance level of 0.05, the same conclusion would be reached, since the p-value would still be greater than α.

Exercises 10.45 – 10.63

10.45 Since this is a two-tailed test, the P-value is equal to twice the area captured in the tail in which z falls. Using Appendix Table 2, the P-values are:

a. 2(0.0179) = 0.0358

b. 2(0.0401) = 0.0802

c. 2(0.2810) = 0.5620

d. $2(0.0749) = 0.1498$

e. $2(0) = 0$

10.47 **a.** P-value = area under the 8 d.f. t curve to the right of 2.0 = 0.040

b. P-value = area under the 13 d.f. t curve to the right of 3.2 = 0.003

c. P-value = area under the 10 d.f. t curve to the left of 2.4

= area under the 10 d.f. t curve to the right of 2.4 = 0.019

d. P-value = area under the 21 d.f. t curve to the left of 4.2

= area under the 21 d.f. t curve to the right of 4.2 = 0.0002

e. P-value = 2(area under the 15 d.f. t curve to the right of 1.6) = 2 (0.065) = 0.13

f. P-value = 2(area under the 15 d.f. t curve to the right of 1.6) = 2 (0.065) = 0.13

g. P-value = 2(area under the 15 d.f. t curve to the right of 6.3) = 2(0) = 0

10.49 The P-value for this test is equal to the area under the 14 d.f. t curve to the right of 3.2 = 0.003.

a.   = 0.05, reject H_o

b.   = 0.01, reject H_o

c.   = 0.001, fail to reject H_o

10.51 **a.** P-value = 2(area under the 12 d.f. t curve to the right of 1.6) = 2(0.068) = 0.136.

Since P-value > , H_o is not rejected.

b. P-value = 2(area under the 12 d.f. t curve to the left of 1.6)

= 2(area under the 12 d.f. t curve to the right of 1.6) = 2(0.068) = 0.136.

Since P-value > , H_o is not rejected.

c. P-value = 2(area under the 24 d.f. t curve to the left of 2.6)

= 2(area under the 24 d.f. t curve to the right of 2.6) = 2(0.008) = 0.016.

Since P-value > , H_o is not rejected.

d. P-value = 2(area under the 24 d.f. t curve to the left of 3.6)

= 2(area under the 24 d.f. t curve to the right of 3.6) = 2(0.001) = 0.002.

H_o would be rejected for any   > 0.002.

10.53 Let   = mean wrist extension while using the new mouse design.

$$H_0 : \mu = 20$$

$$H_a : \mu > 20$$

= 0.05 (A value for   was not specified in the problem. The value 0.05 was chosen for illustration.)

Test statistic: $t = \dfrac{\bar{x} - \text{hypothesized value}}{\dfrac{s}{\sqrt{n}}} = \dfrac{\bar{x} - 20}{\dfrac{s}{\sqrt{n}}}$

This test requires a random sample and either a large sample or a normal population distribution. If this sample was a random sample of Cornell University students, it would be appropriate to generalize the results of the test to the population of Cornell students. However, if we wanted to generalize the results of this study to all university students, we would have to assume that the sample of the 24 Cornell students were a random sample of all university students, which they are clearly not. From the boxplot (below), the data is symmetrical with no outliers, it is not unreasonable to assume that the sample came from a normal distribution.

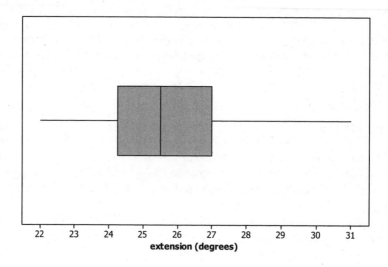

extension (degrees)

$$n = 24, \ \bar{x} = 26.9, \ s = 5.03. \quad t = \frac{26.9 - 20}{\frac{5.03}{\sqrt{24}}} = 6.72$$

P-value = area under the 23 d.f. t curve to the right of 6.72 ≈ 0
Since the P-value is less than , the null hypothesis is rejected at the 0.05 level. The data supports the conclusion that the mean wrist extension for all Cornell undergraduates using this mouse design is greater than 20 degrees.

10.55 A hypothesis test will give one of two results: either statistically significant or not statistically significant; it won't report any quantity or difference. A confidence interval is need for that information. It is possible for a result to be significantly significant but at such a small difference from the null hypothesis that from a practical point of view, the difference has no practical consequence.

10.57 **a.** The large standard deviation means that the distribution for online times for the teens had a lot of variability. As it is large compared to the mean, it is likely that the distribution is heavily skewed to the left.

b. Let μ = mean number of hours per week that teens spend online.

$$H_0 : \mu = 10$$
$$H_a : \mu > 10$$

α = 0.05. (A value for α was not specified in the problem. The value 0.05 was chosen for illustration.)

Test statistic: $t = \dfrac{\bar{x} - \text{hypothesized value}}{\frac{s}{\sqrt{n}}} = \dfrac{\bar{x} - 10}{\frac{s}{\sqrt{n}}}$

This test requires a random sample and either a large sample or a normal population distribution. The teens were randomly selected and although the distribution of the data is probably skewed, because the sample is large ($n = 534$), it is reasonable to proceed with the t test.

$$n = 534, \quad \bar{x} = 14.6, \quad s = 11.6. \quad t = \dfrac{14.6 - 10}{\frac{11.6}{\sqrt{534}}} = 9.16$$

P-value = area under the 533 d.f. t curve to the right of 9.16 $\approx$ 0

Since the P-value is less than α, the null hypothesis is rejected at the 0.05 level. The data supports the conclusion that the mean time spent online by teens is greater than 10 hours per week.

10.59 The authors are saying that the positive effects of music on pain intensity found in the study are statistically significant, but not practically significant from a clinical point of view.

10.61 Population characteristic of interest: μ = population mean MWAL

$$H_o: \mu = 25$$
$$H_a: \mu > 25$$
$$\alpha = 0.05$$

Test statistic: $t = \dfrac{\bar{x} - 25}{\frac{s}{\sqrt{n}}}$

Assumptions: This test requires a random sample and either a large sample size (generally n $\geq$ 30) or a normal population distribution. Since the sample size is only 5, and a box plot of the data does not show perfect symmetry, does this seem reasonable? Based on the their understanding of MWAL values, the authors of the article thought it was reasonable to assume that the population distribution was approximately normal and based on their expert judgment, we will proceed, with caution, with the t-test.

Computations: n = 5, $\bar{x} = 27.54$, s = 5.47

$$t = \dfrac{27.54 - 25}{\frac{5.47}{\sqrt{5}}} = 1.0$$

P-value = area under the 4 d.f. t curve to the right of 1 = 0.187

Conclusion: Since the P-value is greater than α, the null hypothesis is not rejected at the 0.01 level. There is not enough evidence to suggest that the mean MWAL exceeds 25.

10.63   **a.** Since the boxplot is nearly symmetric and the normal probability plot is very much like a straight line, it is reasonable to use a t-test to carry out the hypothesis test on μ.

b. The median is slightly less than 245 and because of the near symmetry, the mean should be close to 245. Also, because of the large amount of variability in the data, it is quite conceivable that the average calorie content is 240.

c. Let μ denote the true average calorie content of this type of frozen dinner.

H_o: $\mu = 240$
H_a: $\mu \neq 240$

$\alpha = 0.05$ (A value for α was not specified in the problem, so this value was chosen). Since the sample is reasonable large (≥ 30) independent and randomly selected, it is reasonable to use the t test.

Test statistic: $t = \dfrac{\bar{x} - 240}{\dfrac{s}{\sqrt{n}}}$ with d.f. = 12 $-$ 1 = 11

Computations: n = 12, $\bar{x}$ = 244.333, s = 12.383

$t = \dfrac{244.333 - 240}{\dfrac{12.383}{\sqrt{12}}} = \dfrac{4.333}{3.575} = 1.21$

P-value = 2(area under the 11 d.f. t curve to the right of 1.21) $\approx$ 2 (0.128) = 0.256 Since the P-value exceeds α, H_o is not rejected. The sample evidence does not support the conclusion that the average calorie content differs from 240.

Exercises 10.65 – 10.71

10.65   **a.** When the significance level is held fixed, increasing the sample size will increase the power of a test.

b. When the sample size is held fixed, increasing the significance level will increase the power of a test.

10.67   **a.** α = area under the z curve to the left of -1.28 = 0.1003.

b. The decision rule is reject H_0 if $\bar{x} < 10 - 1.28(0.1) \Rightarrow \bar{x} < 9.872$.

$z = \dfrac{9.872 - 9.8}{0.1} = 0.72$

β (when μ = 9.8) = area under the z curve to the right of 0.72
$= 1 -$ area under the z curve to the left of 0.72
$= 1 - 0.7642 = 0.2358$.

This means that if μ = 9.8, about 24% of all samples would result in $\bar{x}$ values greater than 9.872 and the nonrejection of H_o : μ = 10.

c. β when μ = 9.5 would be smaller than β when μ = 9.8.

$z = \dfrac{9.872 - 9.5}{0.1} = 3.72$

β (when μ = 9.5) = 1 $-$ area under the z curve to the left of 3.72 $\approx$ 1 $-$ 1 = 0

d. When μ = 9.8, the value of β is 0.2358 from part **b**. So, power of the test = 1- β = 1-0.2358 = 0.7642. When μ = 9.5, the value of β is (approximately) 0 from part **c**. So, power of the test = 1- β = 1.

130

10.69 **a.** Let denote the mean amount of shaft wear after a fixed mileage.

H_o: = 0.035
H_a: > 0.035
 = 0.05

$$t = \frac{\bar{x} - 0.035}{\frac{s}{\sqrt{n}}} \quad \text{with d.f.} = 6$$

$$t = \frac{0.0372 - 0.035}{\frac{0.0125}{\sqrt{7}}} = 0.466$$

P-value = area under the 6 d.f. t curve to the right of 0.466 0.329.
Since the P-value exceeds , H_o is not rejected. It cannot be concluded that the mean amount of shaft wear exceeds 0.035 inches.

b. $d = \dfrac{|\text{alternative value} - \text{hypotheszed value}|}{\sigma} = \dfrac{|0.04 - 0.035|}{0.0125} = 0.4$

From Appendix Table 5, use the set of curves for = 0.05, one-tailed test. Enter the table using d = 0.4, and go up to where the d = 0.4 line intersects with the curve for d.f. = 6. Then read the value of the vertical axis. This leads to a of about 0.75.

c. From **b**, value of is about 0.75. So, power of the test = 1- 1-0.75 = 0.25.

10.71 **a.** $d = \dfrac{|0.52 - 0.5|}{0.02} = 1$ From Appendix Table 5, 0.06.

b. $d = \dfrac{|0.48 - 0.5|}{0.02} = 1$ From Appendix Table 5, 0.06.

c. $d = \dfrac{|0.52 - 0.5|}{0.02} = 1$ From Appendix Table 5, 0.21

d. $d = \dfrac{|0.54 - 0.5|}{0.02} = 2$ From Appendix Table 5, 0.

e. $d = \dfrac{|0.54 - 0.5|}{0.04} = 1$ From Appendix Table 5, 0.06.

f. $d = \dfrac{|0.54 - 0.5|}{0.04} = 1$ From Appendix Table 5, 0.01.

g. Answers will vary from student to student.

Exercises 10.73 – 10.95

10.73 Let π represent the true proportion of all U.S. adults who approve of casino gambling.

$$H_0: \pi = 2/3 = .667$$
$$H_a: \pi > .667$$

$\alpha = 0.05$ (A value for α was not specified in the problem. The value 0.05 was chosen for illustration.)

Test Statistic: $z = \dfrac{p - \text{hypothesized value}}{\sqrt{\dfrac{(\text{hypothesized value})(1 - \text{hypothesized value})}{n}}} = \dfrac{p - .5}{\sqrt{\dfrac{(.667)(.333)}{n}}}$

Assumptions: This test requires a random sample and a large sample size. The given sample was a random sample, the population size is much larger than the sample size, and the sample size was $n = 1523$. Because $1523(.667) \geq 10$ and $1523(.333) \geq 10$, the large-sample test is appropriate.

$n = 1523$, $p = 1035/1523 = .68$.

$$z = \dfrac{0.680 - 0.667}{\sqrt{\dfrac{0.667(0.333)}{1523}}} = 1.08$$

P-value = area under the z curve to the right of 1.08 = .1401

Since the P-value is greater than α, H_0 cannot be rejected.

It would not be all that unusual to observe a sample proportion as large as .680 if the null hypothesis was true. There is not strong evidence that the proportion of all U.S. adults who favor casino gambling is great than two-thirds.

10.75 Let π represent the proportion of APL patients receiving arsenic who go into remission

$$H_0: \pi = 0.15$$
$$H_a: \pi > 0.15$$

We will compute a P-value for this test. $\alpha = 0.01$.

Since $n\pi = 100(0.15) \geq 10$, and $n(1 - \pi) = 100(0.85) \geq 100$, the large sample z test may be used.

$n = 100$, $p = 0.42$

$$z = \dfrac{0.42 - .15}{\sqrt{\dfrac{0.15(0.85)}{100}}} = \dfrac{.27}{.0357} = 7.56$$

P-value = area under the z curve to the right of 7.56 $\approx$ 0

Since the P-value is less than α, H_0 is rejected.

There is enough evidence to suggest that the proportion of APL patients receiving arsenic who go into remission is greater than 0.15

10.77 Let represent the proportion of U.S. adults who are aware that an investment of $25 a week could result in a sum of over $100,000 over 40 years.

H_o: = 0.4

H_a: < 0.4

We will compute a P-value for this test. = 0.05.

Since n = 1010(0.4) 10, and n(1) = 1010(0.6) 10, the large sample z test may be used.

n = 1010, p = 374/1010 = 0.3703

$$z = \frac{0.3703 - .4}{\sqrt{\dfrac{0.4(0.6)}{1010}}} = \frac{-0.0297}{0.0154} = -1.93$$

P-value = area under the z curve to the left of –1.93 0.0268

Since the P-value is smaller than , H_o is rejected.

There is not enough evidence to suggest that proportion of U.S. adults who are aware that an investment of $25 a week could result in a sum of over $100,000 over 40 years is less than 40%.

10.79 Let represent the proportion of social sciences and humanities majors who have a B average going into college but end up with a GPA below 3.0 at the end of their first year.

H_o: = 0.50

H_a: > 0.50

We will compute a P-value for this test. The problem does not specify a value for but we will use = 0.05 for illustration.

Since n = 137(0.5) 10, and n(1) = 137(0.5) 10, the large sample z test may be used.

n = 137, p = 0.532

$$z = \frac{0.532 - 0.5}{\sqrt{\dfrac{0.5(0.5)}{137}}} = \frac{0.032}{0.0427} = 0.7491$$

P-value = area under the z curve to the right of 0.7491= 0.2269.

Since the P-value is greater than , H_o cannot be rejected at the level of significance of 0.05. The data does not support the conclusion that a majority of students majoring in social sciences and humanities who enroll with a HOPE scholarship will lose their scholarship.

10.81 Population characteristic of interest: μ = average age of brides marrying for the first time in 1990.

$$H_o: \mu = 20.8$$
$$H_a: \mu > 20.8$$

$\alpha = 0.01$

Since the sample is reasonable large (≥ 30) independent and randomly selected, it is reasonable to use the t test.

Test statistic: $t = \dfrac{\bar{x} - 20.8}{\dfrac{s}{\sqrt{n}}}$

Computations: n = 100, $\bar{x}$ = 23.9, s = 6.4,

$$t = \frac{23.9 - 20.8}{\dfrac{6.4}{\sqrt{100}}} = \frac{3.10}{0.64} = 4.84$$

P-value = area under the 99 d.f. t curve to the right of 4.84 $\approx 1 - 1 = 0$

Conclusion: Since the P-value is less than α, the null hypothesis is rejected at the 0.01 level. The data supports the conclusion that the mean age of brides marrying for the first time in 1990 is larger than that in 1970.

10.83 Let π represent the proportion of local residents who oppose hunting on Morro Bay

$$H_o: \pi = 0.50$$
$$H_a: \pi > 0.50$$

$\alpha = 0.01$

Since $n\pi = 750(0.50) = 375 \geq 10$, and $n(1 - \pi) = 750(0.5) = 375 \geq 10$, the large sample z test may be used.

n = 750, x = 560, p = $\dfrac{560}{750}$ = 0.7467

$$z = \frac{0.7467 - 0.5}{\sqrt{\dfrac{0.5(0.5)}{750}}} = \frac{0.2467}{0.0183} = 13.51$$

P-value = area under the z curve to the right of 13.51 $\approx 1 - 1 = 0$

Since the P-value is less than α, H_o is rejected.

The data supports the conclusion that the majority of local residents oppose hunting on Morro Bay.

10.85 Let   denote the true proportion of all cars purchased in this area that were white.

H_o: = 0.20
H_a: ≠ 0.20

α = 0.05 and/or 0.01. See the conclusion below
Since n = 400(0.2)   10, and n(1   ) = 400(0.8)   10, the large sample z test may be used.
n = 400, x = 100, p = 0.25

$$z = \dfrac{0.25 - 0.20}{\sqrt{\dfrac{0.2(0.8)}{400}}} = 2.50$$

P-value = 2(area under the z curve to the right of 2.50) = 2(1   0.9938) = 0.0124.
With   = 0.05, the P-value is less than   and the null hypothesis would be rejected. The conclusion would be that the true proportion of cars sold that are white differs from the national rate of 0.20. With   = 0.01, the P-value is greater than   and the null hypothesis would not be rejected. Then the sample does not support the conclusion that the true proportion of cars sold that are white differs from 0.20.

10.87 Let   denote the true average time to change (in months)..

H_o: = 24
H_a: > 24

 = 0.01

Test statistic: $t = \dfrac{\bar{x} - 24}{\dfrac{s}{\sqrt{n}}}$

Computations: n = 44, $\bar{x}$ = 35.02, s = 18.94,

$$t = \dfrac{35.02 - 24}{\dfrac{18.94}{\sqrt{44}}} = 3.86$$

P-value = area under the 43 d.f. t curve to the right of 3.86   1   1 = 0.
Since the P-value is less than , H_o is rejected. There is sufficient information in this sample to support the conclusion that the true average time to change exceeds two years.

10.89 a. Daily caffeine consumption cannot be a negative value. Since the standard deviation is larger than the mean, this would imply that a sizable portion of a normal curve with this mean and this standard deviation would extend into the negative values on the number line. Therefore, it is not plausible that the population distribution of daily caffeine consumption is normal.

Since the sample size is large (greater than 30) the Central Limit Theorem allows for the conclusion that the distribution of $\bar{x}$ is approximately normal even though the population distribution is not normal. So it is not necessary to assume that the population distribution of daily caffeine consumption is normal to test hypotheses about the value of population mean consumption.

b. Let μ denote the population mean daily consumption of caffeine.

$H_o: \mu = 200$

$H_a: \mu > 200$

$\alpha = 0.10$

$$t = \frac{\bar{x} - 200}{\frac{s}{\sqrt{n}}}$$

$n = 47$, $\bar{x} = 215$, $s = 235$

$$t = \frac{\bar{x} - 200}{\frac{s}{\sqrt{n}}} = \frac{215 - 200}{\frac{235}{\sqrt{47}}} = 0.44$$

P-value = area under the 46 d.f. t curve to the right of 0.44 = 1 − 0.6700 = 0.33
Since the P-value exceeds the level of significance of 0.10, H_o is not rejected. The data does not support the conclusion that the population mean daily caffeine consumption exceeds 200 mg.

10.91 Population characteristic of interest: μ = true average fuel efficiency

$H_o: \mu = 30$

$H_a: \mu < 30$

$\alpha = 0.05$ (for demonstration purposes)

Test statistic: $t = \frac{\bar{x} - 30}{\frac{s}{\sqrt{n}}}$

Assumptions: This test requires a random sample and either a large sample size (generally n ≥ 30) or a normal population distribution. Since the sample size is only 6, we can look at a box plot of the data. It shows symmetry, indicating that it would not be unreasonable to assume that the population would be approximately normal. Hence we can proceed with a t-test.

Computations: $n = 6$, $\bar{x} = 29.33$, $s = 1.41$

$$t = \frac{29.33 - 30}{\frac{1.41}{\sqrt{6}}} = -1.164$$

P-value = area under the 5 d.f. t curve to the left of -1.164 = 0.15

Conclusion: Since the P-value is greater than α, the null hypothesis is not rejected at the 0.01 level.

The data does not contradict the prior belief that the true average fuel efficiency is at least 30.

10.93 Let μ denote the true mean time required to achieve $100°$ F with the heating equipment of this manufacturer.

H_o: μ = 15

H_a: μ > 15

α = 0.05

The test statistic is: $t = \dfrac{\bar{x} - 15}{\dfrac{s}{\sqrt{n}}}$.

From the sample: n = 25, $\bar{x}$ = 17.5, s = 2.2,

$t = \dfrac{17.5 - 15}{\dfrac{2.2}{\sqrt{25}}} = 5.68.$

P-value =area under the 24 d.f. t curve to the right of 5.68 $\approx$ 1 $-$ 1 = 0. Because the P-value is smaller than α, H_o is rejected. The data does cast doubt on the company's claim that it requires at most 15 minutes to achieve $100°$ F.

10.95 P(Type I error) = P( ) = 0

P(Type II error) = P( ) = 0.1   P( )   0.3

137

Chapter 11

Exercises 11.1 – 11.27

11.1 $\mu_{\bar{x}_1-\bar{x}_2} = \mu_1 - \mu_2 = 30 - 25 = 5$

$$\sigma_{\bar{x}_1-\bar{x}_2} = \sqrt{\frac{\sigma_1^2}{n_1} + \frac{\sigma_2^2}{n_2}} = \sqrt{\frac{(2)^2}{40} + \frac{(3)^2}{50}} = \sqrt{\frac{4}{40} + \frac{9}{50}} = \sqrt{0.28} = 0.529$$

Since both n_1 and n_2 are large, the sampling distribution of $\bar{x}_1$ $\bar{x}_2$ is approximately normal. It is centered at 5 and the standard deviation is 0.529.

11.3 **a.** The population distributions are approximately normal or the sample size is large and the two samples are independently selected random samples. Since one of the sample sizes is only 22 and we don't have access to the raw data, we must assume that the population distributions are approximately normal.

b. Let μ_1 = mean HRV for all heart attack patients who own dogs.

Let μ_2 = mean HRV for all heart attack patients who do not own dogs.

$$H_0 : \mu_1 - \mu_2 = 0 \qquad H_a : \mu_1 - \mu_2 \neq 0$$
$$= 0.05$$

Test statistic: $t = \dfrac{(\bar{x}_1 - \bar{x}_2) - 0}{\sqrt{\dfrac{s_1^2}{n_1} + \dfrac{s_2^2}{n_2}}}$

Assumptions have been discussed in part (a)

$n_1 = 22$, $\bar{x}_1 = 873$, $s_1 = 136$, $n_2 = 80$, $\bar{x}_2 = 800$, $s_2 = 134$

$$t = \frac{(873 - 800) - 0}{\sqrt{\dfrac{(136)^2}{22} + \dfrac{(134)^2}{80}}} = 2.24$$

$$df = \frac{\left(\dfrac{s_1^2}{n_1} + \dfrac{s_2^2}{n_2}\right)^2}{\dfrac{1}{n_1-1}\left(\dfrac{s_1^2}{n_1}\right)^2 + \dfrac{1}{n_2-1}\left(\dfrac{s_2^2}{n_2}\right)^2} = \frac{1134602.622}{33658.207 + 637.694} = 33.08$$

So $df = 33$ (rounded down to an integer)

P-value = 2(the area under the 39 df t curve to the right of 2.24   2(0.016) = 0.032.Since the P-value is less than , the null hypothesis can be rejected at the 0.05 level of significance. There is enough evidence to show that mean HRV levels in different in heart attack patients who own a dog and who do not own a dog.

11.5 Let μ_1 = mean length of daily commute of a male working adult living in Calgary.

Let μ_2 = mean length of daily commute of a female working adult living in Calgary.

$H_0: \mu_1 - \mu_2 = 0 \qquad H_a: \mu_1 - \mu_2 \neq 0$

$\qquad = 0.05$

Test statistic: $\quad t = \dfrac{(\bar{x}_1 - \bar{x}_2) - 0}{\sqrt{\dfrac{s_1^2}{n_1} + \dfrac{s_2^2}{n_2}}}$

The sample sizes are large (247 and 253) and it is stated that the samples are independently selected random samples.

$n_1 = 247,\ \bar{x}_1 = 29.6,\ s_1 = 24.3,\ n_2 = 253,\ \bar{x}_2 = 27.3,\ s_2 = 24.0$

$t = \dfrac{(29.6 - 27.3) - 0}{\sqrt{\dfrac{(24.3)^2}{247} + \dfrac{(24)^2}{253}}} = 1.06$

$df = \dfrac{\left(\dfrac{s_1^2}{n_1} + \dfrac{s_2^2}{n_2}\right)^2}{\dfrac{1}{n_1 - 1}\left(\dfrac{s_1^2}{n_1}\right)^2 + \dfrac{1}{n_2 - 1}\left(\dfrac{s_2^2}{n_2}\right)^2} = \dfrac{21.7839}{0.02323 + 0.02057} = 497.36$

So $df = 497$ (rounded down to an integer)

P-value = 2(the area under the 497 df t curve to the right of 1.06) 2(0.143) = 0.2876
Since the P-value is greater than , the null hypothesis cannot be rejected at the 0.05 level of significance. There is not enough evidence to show that there is a difference in mean commute times for male and female working Calgary residents.

11.7 **a.** $H_0: \mu_1 = \mu_2$ vs. $H_a: \mu_1 < \mu_2$, where μ_1 is the mean payment for all claims not involving errors and μ_2 is the mean payment for all claims involving errors.

b. With sample sizes so large (515 and 889) the df will be large, and the bottom row of the t table can be used. With a P-value of 0.004, and a lower-tailed test, the value of the test statistic must have been between 2.58 and 3.09. t = 2.65 is the best answer.

11.9 **a.** Let μ_1 denote the true mean breaking force in a dry medium at 37 degrees and μ_2 denote the true mean breaking force in a wet medium at 37 degrees. Although the distribution of the sample taken from both conditions are skewed, there are no outliers and it would not be unreasonable to assume they both came from approximately normal populations. We would have to assume the samples were independent and taken at random.

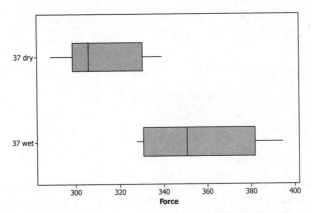

$n_1 = 6$, $\bar{x}_1 = 311.6$, $s_1 = 18.4$, $n_2 = 6$, $\bar{x}_2 = 355.6$, $s_2 = 27.3$.

$$df = \frac{\left(\dfrac{s_1^2}{n_1} + \dfrac{s_2^2}{n_2}\right)^2}{\dfrac{1}{n_1-1}\left(\dfrac{s_1^2}{n_1}\right)^2 + \dfrac{1}{n_2-1}\left(\dfrac{s_2^2}{n_2}\right)^2} = \frac{32631.424}{636.79 + 3085.873} = 8.765$$

df = 8 so t critical value =1.86.

Confidence level: 90%:

$$(311.6 - 355.6) \pm 1.86\sqrt{\frac{(18.4)^2}{6} + \frac{(27.3)^2}{6}}$$

$$\Rightarrow\ -44 \pm 1.86(13.44) \ \Rightarrow\ -44 \pm 25.0 \ \Rightarrow\ (-69,\ -19)$$

Based on this sample, we are highly confident that $\mu_1 - \mu_2$ is between -69 and -19. Therefore, with 90% confidence, we believe that the mean breaking force required to break a cement bond is between 19 and 69 Newtons greater in a wet medium than a dry medium when the temperature is 37 degrees.

b. Let μ_1 denote the true mean breaking force in a dry medium at 37 degrees and μ_2 denote the true mean breaking force in a dry medium at 22 degrees.

$$H_0 : \mu_1 - \mu_2 = 100 \qquad H_a : \mu_1 - \mu_2 > 100$$
$$= 0.10$$

Test statistic: $t = \dfrac{(\bar{x}_1 - \bar{x}_2) - 100}{\sqrt{\dfrac{s_1^2}{n_1} + \dfrac{s_2^2}{n_2}}}$

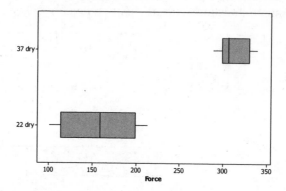

Although the distribution of the sample taken from both conditions are skewed, there are no outliers and it would not be unreasonable to assume they both came from approximately normal populations. We would have to assume the samples were independent and taken at random.

$n_1 = 6$, $\bar{x}_1 = 311.6$, $s_1 = 18.4$, $n_2 = 6$, $\bar{x}_2 = 157.5$, $s_2 = 44.3$

$t = \dfrac{(311.6 - 157.5) - 100}{\sqrt{\dfrac{(18.4)^2}{6} + \dfrac{(44.3)^2}{6}}} = 2.76$

$df = \dfrac{\left(\dfrac{s_1^2}{n_1} + \dfrac{s_2^2}{n_2}\right)^2}{\dfrac{1}{n_1 - 1}\left(\dfrac{s_1^2}{n_1}\right)^2 + \dfrac{1}{n_2 - 1}\left(\dfrac{s_2^2}{n_2}\right)^2} = \dfrac{147081.20}{636.64 + 21396.27} = 6.75$

So $df = 6$ (rounded down to an integer)

P-value = the area under the 6 df t curve to the right of 2.76   0.016

Since the P-value is less than , the null hypothesis can be rejected at the 0.05 level of significance. There is enough evidence to show that the mean breaking force in a dry medium at the higher temperature is greater than the mean breaking force at the lower temperature by more than 100 N.

11.11 **a.** Let $_1$ be the true mean percentage of time male monkeys spend playing with a police car and $_2$ be the true mean percentage of the time female monkeys spend playing with a police car.

H_o: $\mu_1 - \mu_2 = 0$ H_a: $\mu_1 - \mu_2 > 0$

 = 0.05

Test statistic: $t = \dfrac{(\bar{x}_1 - \bar{x}_2) - 0}{\sqrt{\dfrac{s_1^2}{n_1} + \dfrac{s_2^2}{n_2}}}$

The sample sizes for the two groups are large (both 44) and we can assume that the two samples are representative of the populations of male and female monkeys so we can regard them as random samples.

$n_1 = 44$, $\bar{x}_1 = 18$, $s_1 = 5$, $n_2 = 44$, $\bar{x}_2 = 8$, $s_2 = 4$

$t = \dfrac{(18 - 8) - 0}{\sqrt{\dfrac{(5)^2}{44} + \dfrac{(4)^2}{44}}} = 10.4$

$df = \dfrac{\left(\dfrac{s_1^2}{n_1} + \dfrac{s_2^2}{n_2}\right)^2}{\dfrac{1}{n_1 - 1}\left(\dfrac{s_1^2}{n_1}\right)^2 + \dfrac{1}{n_2 - 1}\left(\dfrac{s_2^2}{n_2}\right)^2} = \dfrac{.8678}{0.0075 + .0031} = 82.01$

So $df = 82$ (rounded down to an integer).

P-value = the area under the 82 df t curve to the right of 10.4 ≈ 0.0000.
Since the P-value is much less than   , the null hypothesis of no difference is rejected. There is very strong evidence to indicate that that the mean percentage of the time spent playing with the police car is greater for male monkeys than for female monkeys.

b. Let $_1$ be the true mean percentage of time female monkeys spend playing with a doll and $_2$ be the true mean percentage of the time male monkeys spend playing with a doll.

H_o: $\mu_1 - \mu_2 = 0$ H_a: $\mu_1 - \mu_2 > 0$

  = 0.05

Test statistic: $t = \dfrac{(\bar{x}_1 - \bar{x}_2) - 0}{\sqrt{\dfrac{s_1^2}{n_1} + \dfrac{s_2^2}{n_2}}}$

The sample sizes for the two groups are large (both 44) and we can assume that the two samples are representative of the populations of male and female monkeys so we can regard them as random samples.

$n_1 = 44$, $\bar{x}_1 = 20$, $s_1 = 4$, $n_2 = 44$, $\bar{x}_2 = 9$, $s_2 = 2$

$t = \dfrac{(20 - 9) - 0}{\sqrt{\dfrac{(4)^2}{44} + \dfrac{(2)^2}{44}}} = 16.3$

$$df = \dfrac{\left(\dfrac{s_1^2}{n_1} + \dfrac{s_2^2}{n_2}\right)^2}{\dfrac{1}{n_1-1}\left(\dfrac{s_1^2}{n_1}\right)^2 + \dfrac{1}{n_2-1}\left(\dfrac{s_2^2}{n_2}\right)^2} = \dfrac{.2069}{.0031 + .00019} = 62.88$$

So $df = 62$ (rounded down to an integer)

P-value = the area under the 62 df t curve to the right of 16.3 ≈ 0.0000.

Since the P-value is much less than , the null hypothesis can be rejected. There is very strong evidence to indicate that that the mean percentage of the time spent playing with the doll is greater for female monkeys than for male monkeys.

c. Let $_1$ be the true mean percentage of time female monkeys spend playing with a furry dog and $_2$ be the true mean percentage of the time male monkeys spend playing with a furry dog.

H_o: $\mu_1 - \mu_2 = 0$ H_a: $\mu_1 - \mu_2 \neq 0$

= 0.05

Test statistic: $t = \dfrac{(\overline{x}_1 - \overline{x}_2) - 0}{\sqrt{\dfrac{s_1^2}{n_1} + \dfrac{s_2^2}{n_2}}}$

The sample sizes for the two groups are large (both 44) and we can assume that the two samples are representative of the populations of male and female monkeys so we can regard them as random samples.

$n_1 = 44$, $\overline{x}_1 = 20$, $s_1 = 5$, $n_2 = 44$, $\overline{x}_2 = 25$, $s_2 = 5$

$$t = \dfrac{(20 - 25) - 0}{\sqrt{\dfrac{(5)^2}{44} + \dfrac{(5)^2}{44}}} = -4.69$$

$$df = \dfrac{\left(\dfrac{s_1^2}{n_1} + \dfrac{s_2^2}{n_2}\right)^2}{\dfrac{1}{n_1-1}\left(\dfrac{s_1^2}{n_1}\right)^2 + \dfrac{1}{n_2-1}\left(\dfrac{s_2^2}{n_2}\right)^2} = \dfrac{1.29}{.0075 + .0075} = 85.96$$

So $df = 85$ (rounded down to an integer)

P-value = 2(the area under the 85 df t curve to the left of -4.69) $\approx 2(0.00) \approx 0.00$.

Since the P-value is much less than , the null hypothesis of no difference is rejected. There is very strong evidence to indicate that that the mean percentage of the time spent playing with the furry dog differs between male and female monkeys.

d. Although it appears that the male monkeys spent time playing with the "boy" toys and the female monkeys spent more time playing with the "girl" toys, there was a difference in the "neutral" toy as well. Any difference in the time spent playing with these toys may have nothing to do with the gender choices, males monkeys may like a certain color better, or shiny surfaces or some other underlying factor. This is an observational study and as such, causation cannot be shown.

e. One of the conditions of a two-sample t test is that the two samples are independent. In this example, if the mean percentage of the time spent playing with the police car increased, the mean percentage of the time spent playing with the doll would have to decrease. Therefore one affects the other, they are not independent and this violates one of the conditions of the test.

11.13 μ_1 be the mean Wechsler Memory Scale score taken by the Ginko group

μ_2 be the mean Wechsler Memory Scale score taken by the control group

$H_o:\ \mu_1 - \mu_2 = 0$ $H_a:\ \mu_1 - \mu_2 > 0$

$\alpha = 0.05$

Test statistic: $t = \dfrac{(\bar{x}_1 - \bar{x}_2) - 0}{\sqrt{\dfrac{s_1^2}{n_1} + \dfrac{s_2^2}{n_2}}}$

Assumptions: The population distributions are approximately normal or the sample size is large (generally $\geq$ 30) and the two samples are independently selected random samples.

$n_1 = 104,\ \bar{x}_1 = 5.6,\ s_1 = 0.6,\ n_2 = 115,\ \bar{x}_2 = 5.5,\ s_2 = 0.6$

$t = \dfrac{(5.6 - 5.5) - 0}{\sqrt{\dfrac{(0.6)^2}{104} + \dfrac{(0.6)^2}{115}}} = \dfrac{0.1}{0.0812} = 1.23$

$df = \dfrac{\left(\dfrac{s_1^2}{n_1} + \dfrac{s_2^2}{n_2}\right)^2}{\dfrac{1}{n_1 - 1}\left(\dfrac{s_1^2}{n_1}\right)^2 + \dfrac{1}{n_2 - 1}\left(\dfrac{s_2^2}{n_2}\right)^2} = \dfrac{(0.0035 + 0.0031)^2}{\dfrac{(0.0035)^2}{103} + \dfrac{(0.0031)^2}{114}} = 214.34$

So $df = 214$ (rounded down to an integer)

P-value = the area under the 214 df t curve to the right of 1.23 $\approx$ 0.110

Since the P-value is greater than α, the null hypothesis is not rejected at the 0.05 level of significance. There is not enough evidence to show that taking 40 mg ginkgo three times a day is effective in increasing mean performance on the Wechsler Memory Scale.

11.15 **a.** Let μ_1 be the true mean "appropriateness" score assigned to wearing a hat in a class by the population of students and μ_2 be the corresponding score for faculty.

$H_o: \mu_1 - \mu_2 = 0 \quad H_a: \mu_1 - \mu_2 \neq 0$

$\alpha = 0.05$ (A value for α is not specified in the problem. We use $\alpha = 0.05$ for illustration.)

Test statistic: $t = \dfrac{(\bar{x}_1 - \bar{x}_2) - 0}{\sqrt{\dfrac{s_1^2}{n_1} + \dfrac{s_2^2}{n_2}}}$

Assumptions: The sample sizes for the two groups are large (say, greater than 30 for each) and the two samples are independently selected random samples.

$n_1 = 173, \ \bar{x}_1 = 2.80, \ s_1 = 1.0, \ n_2 = 98, \ \bar{x}_2 = 3.63, \ s_2 = 1.0$

$t = \dfrac{(2.80 - 3.63) - 0}{\sqrt{\dfrac{(1.0)^2}{173} + \dfrac{(1.0)^2}{98}}} = \dfrac{-0.83}{0.1264} = -6.5649$

$df = \dfrac{\left(\dfrac{s_1^2}{n_1} + \dfrac{s_2^2}{n_2}\right)^2}{\dfrac{1}{n_1 - 1}\left(\dfrac{s_1^2}{n_1}\right)^2 + \dfrac{1}{n_2 - 1}\left(\dfrac{s_2^2}{n_2}\right)^2} = \dfrac{(0.00578 + 0.01020)^2}{\dfrac{(0.00578)^2}{172} + \dfrac{(0.01020)^2}{97}} = 201.5$

So $df = 201$ (rounded down to an integer)

P-value = 2 times the area under the 201 df t curve to the left of $-6.5649 \approx 0.0000$. Since the P-value is much less than α, the null hypothesis of no difference is rejected. The data do provide very strong evidence to indicate that there is a difference in the mean appropriateness scores between students and faculty for wearing hats in the class room. The mean appropriateness score for students is significantly smaller than that for faculty.

b. Let μ_1 be the true mean "appropriateness" score assigned to addressing an instructor by his or her first name by the population of students and μ_2 be the corresponding score for faculty.

$H_o: \mu_1 - \mu_2 = 0 \quad H_a: \mu_1 - \mu_2 > 0$

$\alpha = 0.05$ (A value for α is not specified in the problem. We use $\alpha = 0.05$ for illustration.)

Test statistic: $t = \dfrac{(\bar{x}_1 - \bar{x}_2) - 0}{\sqrt{\dfrac{s_1^2}{n_1} + \dfrac{s_2^2}{n_2}}}$

Assumptions: The sample sizes for the two groups are large (say, greater than 30 for each) and the two samples are independently selected random samples.

$n_1 = 173, \ \bar{x}_1 = 2.90, \ s_1 = 1.0, \ n_2 = 98, \ \bar{x}_2 = 2.11, \ s_2 = 1.0$

$t = \dfrac{(2.90 - 2.11) - 0}{\sqrt{\dfrac{(1.0)^2}{173} + \dfrac{(1.0)^2}{98}}} = \dfrac{0.79}{0.1264} = 6.2485$

$$df = \frac{\left(\dfrac{s_1^2}{n_1} + \dfrac{s_2^2}{n_2}\right)^2}{\dfrac{1}{n_1-1}\left(\dfrac{s_1^2}{n_1}\right)^2 + \dfrac{1}{n_2-1}\left(\dfrac{s_2^2}{n_2}\right)^2} = \frac{(0.00578+0.01020)^2}{\dfrac{(0.00578)^2}{172} + \dfrac{(0.01020)^2}{97}} = 201.5$$

So $df = 201$ (rounded down to an integer)

P-value = the area under the 201 df t curve to the right of 6.2485 ≈ 0.0000.
Since the P-value is much less than , the null hypothesis of no difference is rejected. The data do provide very strong evidence to indicate that the mean appropriateness score for addressing the instructor by his or her first name is higher for students than for faculty.

c. Let $_1$ be the true mean "appropriateness" score assigned to talking on a cell phone during class by the population of students and $_2$ be the corresponding score for faculty.

H_o: $\mu_1 - \mu_2 = 0$ H_a: $\mu_1 - \mu_2 \neq 0$

= 0.05 (A value for is not specified in the problem. We use =0.05 for illustration.)

Test statistic: $t = \dfrac{(\bar{x}_1 - \bar{x}_2) - 0}{\sqrt{\dfrac{s_1^2}{n_1} + \dfrac{s_2^2}{n_2}}}$

Assumptions: The sample sizes for the two groups are large (say, greater than 30 for each) and the two samples are independently selected random samples.

$n_1 = 173$, $\bar{x}_1 = 1.11$, $s_1 = 1.0$, $n_2 = 98$, $\bar{x}_2 = 1.10$, $s_2 = 1.0$

$$t = \frac{(1.11-1.10)-0}{\sqrt{\dfrac{(1.0)^2}{173} + \dfrac{(1.0)^2}{98}}} = \frac{-0.01}{0.1264} = 0.0791$$

$$df = \frac{\left(\dfrac{s_1^2}{n_1} + \dfrac{s_2^2}{n_2}\right)^2}{\dfrac{1}{n_1-1}\left(\dfrac{s_1^2}{n_1}\right)^2 + \dfrac{1}{n_2-1}\left(\dfrac{s_2^2}{n_2}\right)^2} = \frac{(0.00578+0.01020)^2}{\dfrac{(0.00578)^2}{172} + \dfrac{(0.01020)^2}{97}} = 201.5$$

So $df = 201$ (rounded down to an integer)

P-value = 2 times the area under the 201 df t curve to the right of 0.0791 ≈ 0.9370.
Since the P-value is not less than , the null hypothesis of no difference cannot be rejected. The data do not provide evidence to indicate that there is a difference in the mean appropriateness scores between students and faculty for talking on cell phones in class. The result does not imply that students and faculty consider it acceptable to talk on a cell phone during class. It simply says that data do not provide enough evidence to claim a difference exists.

11.17 Let μ_1 denote the true mean "intention to take science courses" score for male students and μ_2 be the corresponding score for female students. Then $\mu_1 - \mu_2$ denotes the difference between the means of the intention scores for males and females.

$$n_1 = 203, \overline{x}_1 = 3.42, s_1 = 1.49, n_2 = 224, \overline{x}_2 = 2.42, s_2 = 1.35$$

$$V_1 = \frac{s_1^2}{n_1} = \frac{(1.49)^2}{203} = 0.01094 \qquad V_2 = \frac{s_2^2}{n_2} = \frac{(1.35)^2}{224} = 0.00814$$

$$df = \frac{(V_1 + V_2)^2}{\dfrac{V_1^2}{n_1 - 1} + \dfrac{V_2^2}{n_2 - 1}} = \frac{(0.01094 + 0.00814)^2}{\dfrac{(0.01094)^2}{202} + \dfrac{(0.00814)^2}{223}} = 409.2$$

Use df = 409. The *t critical value* is 2.588.

The 95% confidence interval for $\mu_1 - \mu_2$ based on this sample is

$$(3.42 - 2.42) \pm 2.588\sqrt{0.01094 + 0.00814} \implies 1 \pm 2.588(0.13811) \implies (0.6426, 1.3574).$$

Observe that the interval does not include 0, and so 0 is not one of the plausible values of $\mu_1 - \mu_2$. As a matter of fact, the plausible values are in the interval from 0.6426 to 1.3574. The data provide sufficient evidence to conclude that the mean "intention to take science courses" scores for male students is greater than that for female students.

11.19 **a.** Let μ_1 denote the true mean hardness for chicken chilled 0 hours before cooking and μ_2 the true mean hardness for chicken chilled 2 hours before cooking.

$$H_0: \mu_1 - \mu_2 = 0 \qquad H_a: \mu_1 - \mu_2 \neq 0$$
$$\alpha = 0.05$$

Test statistic: $\quad t = \dfrac{(\overline{x}_1 - \overline{x}_2) - 0}{\sqrt{\dfrac{s_1^2}{n_1} + \dfrac{s_2^2}{n_2}}}$

Assumptions: The sample sizes for each group is large (greater than or equal to 30) and the two samples are independently selected random samples.

$$n_1 = 36, \overline{x}_1 = 7.52, s_1 = 0.96, n_2 = 36, \overline{x}_2 = 6.55, s_2 = 1.74$$

$$t = \frac{(7.52 - 6.55) - 0}{\sqrt{\dfrac{(0.96)^2}{36} + \dfrac{(1.74)^2}{36}}} = \frac{0.97}{0.33121} = 2.93$$

$$df = \frac{\left(\dfrac{s_1^2}{n_1} + \dfrac{s_2^2}{n_2}\right)^2}{\dfrac{1}{n_1 - 1}\left(\dfrac{s_1^2}{n_1}\right)^2 + \dfrac{1}{n_2 - 1}\left(\dfrac{s_2^2}{n_2}\right)^2} = \frac{(0.0256 + 0.0841)^2}{\dfrac{(0.0256)^2}{35} + \dfrac{(0.0841)^2}{35}} = 54.5$$

So *df* = 54 (rounded down to an integer)

P-value = 2(area under the 54 df t curve to the right of 2.93) = 2(1 − 0.9975) = 2(0.00249) = 0.00498.

Since the P-value is less than , the null hypothesis is rejected. At level of significance 0.05, there is sufficient evidence to conclude that there is a difference in mean hardness of chicken chilled 0 hours before cooking and chicken chilled 2 hours before cooking.

b. Let $_1$ denote the true mean hardness for chicken chilled 8 hours before cooking and $_2$ the true mean hardness for chicken chilled 24 hours before cooking.

$$H_o: \mu_1 - \mu_2 = 0 \quad H_a: \mu_1 - \mu_2 \neq 0$$

$$= 0.05$$

Test statistic: $t = \dfrac{(\bar{x}_1 - \bar{x}_2) - 0}{\sqrt{\dfrac{s_1^2}{n_1} + \dfrac{s_2^2}{n_2}}}$

Assumptions: The sample sizes for each group is large (greater than or equal to 30) and the two samples are independently selected random samples.

$n_1 = 36$, $\bar{x}_1 = 5.70$, $s_1 = 1.32$, $n_2 = 36$, $\bar{x}_2 = 5.65$, $s_2 = 1.50$

$$t = \frac{(5.70 - 5.65) - 0}{\sqrt{\dfrac{(1.32)^2}{36} + \dfrac{(1.50)^2}{36}}} = \frac{0.05}{0.333017} = 0.15$$

$$df = \frac{\left(\dfrac{s_1^2}{n_1} + \dfrac{s_2^2}{n_2}\right)^2}{\dfrac{1}{n_1 - 1}\left(\dfrac{s_1^2}{n_1}\right)^2 + \dfrac{1}{n_2 - 1}\left(\dfrac{s_2^2}{n_2}\right)^2} = \frac{(0.0484 + 0.0625)^2}{\dfrac{(0.0484)^2}{35} + \dfrac{(0.0625)^2}{35}} = 68.9$$

So $df = 68$ (rounded down to an integer)

P-value = 2(area under the 68 df t curve to the right of 0.15) = 2(1 − 0.5595)
= 2(0.44055) = 0.8811.

Since the P-value exceeds , the null hypothesis is not rejected. At level of significance 0.05, there is not sufficient evidence to conclude that there is a difference in mean hardness of chicken chilled 8 hours before cooking and chicken chilled 24 hours before cooking.

c. Let $_1$ denote the true mean hardness for chicken chilled 2 hours before cooking and $_2$ the true mean hardness for chicken chilled 8 hours before cooking.

$n_1 = 36$, $\bar{x}_1 = 6.55$, $s_1 = 1.74$, $n_2 = 36$, $\bar{x}_2 = 5.70$, $s_2 = 1.32$

$$(6.55 - 5.70) \pm 1.669\sqrt{\dfrac{(1.74)^2}{36} + \dfrac{(1.32)^2}{36}}$$

$\Rightarrow .85 \pm 1.669(.364005) \Rightarrow .85 \pm .6075 \Rightarrow (.242, 1.458)$

Based on this sample, we believe that the mean hardness for chicken chilled for 2 hours before cooking is larger than the mean hardness for chicken chilled 8 hours before cooking. The difference may be as small as 0.242, or may be as large as 1.458.

11.21 Let μ_1 denote the true mean alkalinity for upstream locations and μ_2 the true mean alkalinity for downstream locations.

$H_0: \mu_1 - \mu_2 = -50$ $H_a: \mu_1 - \mu_2 < -50$

$\alpha = 0.05$

Test statistic: $t = \dfrac{(\bar{x}_1 - \bar{x}_2) - (-50)}{\sqrt{\dfrac{s_1^2}{n_1} + \dfrac{s_2^2}{n_2}}}$

Assumptions: The distribution of alkalinity is approximately normal for both types of sites (upstream and downstream) and the two samples are independently selected random samples.

$n_1 = 24$, $\bar{x}_1 = 75.9$, $s_1 = 1.83$, $n_2 = 24$, $\bar{x}_2 = 183.6$, $s_2 = 1.70$

$t = \dfrac{(75.9 - 183.6) - (-50)}{\sqrt{\dfrac{(1.83)^2}{24} + \dfrac{(1.70)^2}{24}}} = \dfrac{-57.7}{0.50986} = 113.17$

$df = \dfrac{\left(\dfrac{s_1^2}{n_1} + \dfrac{s_2^2}{n_2}\right)^2}{\dfrac{1}{n_1 - 1}\left(\dfrac{s_1^2}{n_1}\right)^2 + \dfrac{1}{n_2 - 1}\left(\dfrac{s_2^2}{n_2}\right)^2} = \dfrac{(0.1395 + 0.1204)^2}{\dfrac{(0.1395)^2}{23} + \dfrac{(0.1204)^2}{23}} = 45.75$

So $df = 45$ (rounded down to an integer)

P-value = area under the 45 df t curve to the right of 113 $\approx$ 0.

Since the P-value is less than α, the null hypothesis is rejected. The data supports the conclusion that the true mean alkalinity score for downstream sites is more than 50 units higher than that for upstream sites.

11.23 Let μ_1 denote the mean frequency of alcohol use for those that rush a sorority and μ_2 denote the mean frequency of alcohol use for those that do not rush a sorority.

$H_0: \mu_1 - \mu_2 = 0$ $H_a: \mu_1 - \mu_2 > 0$

$\alpha = 0.01$

Test statistic: $t = \dfrac{(\bar{x}_1 - \bar{x}_2) - 0}{\sqrt{\dfrac{s_1^2}{n_1} + \dfrac{s_2^2}{n_2}}}$

Assumptions: The sample size for each group is large (greater than or equal to 30) and the two samples are independently selected random samples.

$n_1 = 54$, $\bar{x}_1 = 2.72$, $s_1 = 0.86$, $n_2 = 51$, $\bar{x}_2 = 2.11$, $s_2 = 1.02$

$t = \dfrac{(2.72 - 2.11) - 0}{\sqrt{\dfrac{(0.86)^2}{54} + \dfrac{(1.02)^2}{51}}} = \dfrac{0.61}{0.184652} = 3.30$

$$df = \frac{\left(\dfrac{s_1^2}{n_1} + \dfrac{s_2^2}{n_2}\right)^2}{\dfrac{1}{n_1-1}\left(\dfrac{s_1^2}{n_1}\right)^2 + \dfrac{1}{n_2-1}\left(\dfrac{s_2^2}{n_2}\right)^2} = \frac{(0.0137+0.0204)^2}{\dfrac{(0.0137)^2}{53} + \dfrac{(0.0204)^2}{50}} = 98.002$$

So df = 98 (rounded down to an integer)

P-value = area under the 98 df t curve to the right of 3.30 = 1 − 0.9993 = 0.0007

Since the P-value is less than , the null hypothesis is rejected. The data supports the conclusion that the true mean frequency of alcohol use is larger for those that rushed a sorority than for those who did not rush a sorority.

11.25 Let μ_1 denote the mean half-life of vitamin D in plasma for people on a normal diet. Let μ_2 denote the mean half-life of vitamin D in plasma for people on a high-fiber diet. Let $\mu_1 - \mu_2$ denote the true difference in mean half-life of vitamin D in plasma for people in these two groups (normal minus high fiber).

H_o: $\mu_1 - \mu_2 = 0$ H_a: $\mu_1 - \mu_2 > 0$

   = 0.01

Test statistic: $\quad t = \dfrac{(\bar{x}_1 - \bar{x}_2) - 0}{\sqrt{\dfrac{s_1^2}{n_1} + \dfrac{s_2^2}{n_2}}}$

Assumptions: The population distributions are (at least approximately) normal and the two samples are independently selected random samples.

Refer to the Minitab output given in the problem statement.

From the Minitab output the P-value = 0.007. Since the P-value is less than , H_o is rejected. There is sufficient evidence to conclude that the mean half-life of vitamin D is longer for those on a normal diet than for those on a high-fiber diet.

11.27 Let μ_1 denote the mean self-esteem score for students classified as having short duration loneliness. Let μ_2 denote the mean self-esteem score for students classified as having long duration loneliness.

H_o: $\mu_1 - \mu_2 = 0$ H_a: $\mu_1 - \mu_2 > 0$

   = 0.01

Test statistic: $\quad t = \dfrac{(\bar{x}_1 - \bar{x}_2) - 0}{\sqrt{\dfrac{s_1^2}{n_1} + \dfrac{s_2^2}{n_2}}}$

Assumptions: The population distributions are (at least approximately) normal and the two samples are independently selected random samples.

n_1 72, $\overline{x}_1$ 76.78 , s_1 17.8, n_2 17, $\overline{x}_2$ 64.00, s_2 15.68

$$t = \frac{(76.78 - 64.00) - 0}{\sqrt{\frac{(17.8)^2}{72} + \frac{(15.68)^2}{17}}} = \frac{12.78}{4.34316} = 2.9426$$

$$df = \frac{\left(\frac{s_1^2}{n_1} + \frac{s_2^2}{n_2}\right)^2}{\frac{1}{n_1 - 1}\left(\frac{s_1^2}{n_1}\right)^2 + \frac{1}{n_2 - 1}\left(\frac{s_2^2}{n_2}\right)^2} = \frac{(4.4006 + 14.4625)^2}{\frac{(4.4006)^2}{71} + \frac{(14.4625)^2}{16}} = 26.7$$

So df = 26 (rounded down to an integer)

P-value = area under the 26 df t curve to the right of 2.9426   0.0034.

Since the P-value is less than   , H_o is rejected. The sample data supports the conclusion that the mean self esteem is lower for students classified as having long duration loneliness than for students classified as having short duration loneliness.

Exercises 11.29 – 11.43

11.29   **a.** If possible, treat each patient with both drugs with one drug used on one eye and the other drug used on the other eye. For each patient, determine at random which eye will receive the new treatment. Then take observations (readings) of eye pressure on each eye. If this treatment method is not possible, then request the ophthalmologist to pair patients according to their eye pressure so that the two people in a pair have approximately equal eye pressure. Then select one patient from each pair to receive the new drug and treat the other patient in each pair with the standard treatment. Record the reduction in eye pressure. Treat the other person in that pair with the standard treatment and record the reduction in eye pressure. These two readings would constitute a pair. Repeat for each of the other pairs to obtain the paired sample data.

b. Both procedures above would result in paired data.

c. Assign subjects at random to one of the two treatment groups. Measure reduction in eye pressure for both groups. The resulting observations would constitute independent samples.

This experiment is probably not as informative as a paired experiment with the same number of subjects to patient to patient variability which can be quite large.

11.31 Let d = the difference in the time to exhaustion after chocolate milk and after carbohydrate replacement drink., i.e. time after chocolate milk – time after carbohydrate replacement drink.

Let μ_d denote the mean difference in the time to exhaustion.

$H_o: \mu_d = 0 \qquad H_a: \mu_d > 0$

$= 0.05$

We must assume that the sample is randomly selected. The sample size is small, but a boxplot of the differences shows a reasonably symmetrical shape with no outliers. It would not be unreasonable to assume that the sample data comes from a population that is approximately normal.

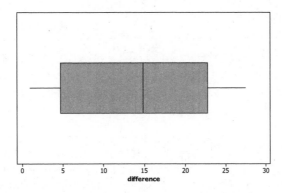

The test statistic is: $t = \dfrac{\overline{x}_d - 0}{\dfrac{s_d}{\sqrt{n}}}$ with d.f. = 8

$\overline{d} = 14.0789$ and $s_d = 9.4745$ $\qquad t = \dfrac{14.0789 - 0}{\dfrac{9.4745}{\sqrt{9}}} = 4.46$

P-value = the area under the 8 df t curve to the right of 4.46 0.00.

Since the P-value is much less than , the null hypothesis should be rejected at 0.05 There is sufficient evidence to suggest that the mean time to exhaustion is greater after chocolate milk than after carbohydrate replacement drink.

11.33 Let d = the difference in total body mineral content for each mother between breast feeding (B) and post weaning. (P), i.e. P - B.

Let μ_d denote the mean difference for all women,

$H_o: \mu_d = 25 \qquad H_a: \mu_d > 25$

$= 0.05$

Assumptions: The sample of mothers are randomly selected. The sample size is small, and the boxplot of the differences (below) shows a skewed distribution, but it has no outliers. It would not be unreasonable to assume that the sample data comes from a population that is approximately normal

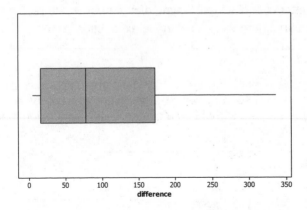

The test statistic is: $t = \dfrac{\bar{x}_d - 25}{\dfrac{s_d}{\sqrt{n}}}$ with d.f. = 9

$\bar{d}$ = 105.7 and s_d = 103.845 $t = \dfrac{105.7 - 25}{\dfrac{103.845}{\sqrt{10}}} = 2.46$

P-value = the area under the 9 df t curve to the right of 2.46   .017

Since the P-value is less than , the null hypothesis should be rejected at 0.05 There is sufficient evidence to suggest that the true average body bone mineral content during post weaning exceeds that during breast feeding by more than 25 grams.

11.35 **a.** Let $_d$ denote the true average difference in translation between dominant and nondominant arms for pitchers (dominant – nondominant)
We assume the pitchers were chosen as random and as the same pitcher is used for both arms, the data is paired. The sample size is small but the boxplot of the differences shows a fairly symmetrical distribution with one outlier. It is not unreasonable to assume that the data comes from a population that is approximately normal.

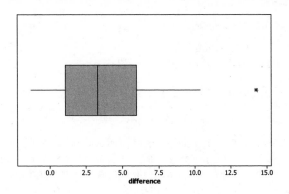

$\bar{d}$ =4.066, s_d = 3.955, n =17.

The 95% confidence interval for μ_d with df =16 is:

$$\bar{d} \pm (t \text{ critical})\frac{s_d}{\sqrt{n}} \implies 4.066 \pm (2.12)\left(\frac{3.955}{\sqrt{17}}\right)$$

$$\implies 4.066 \pm 2.034 \implies (2.032, 6.10).$$

With 95% confidence, it is estimated that pitchers using their dominant arm have, a greater average difference in the translation of their shoulder than when they use their nondominant arm by between 2 032 mm and 6.10 mm.

b. Let $_d$ denote the true average difference in translation between dominant and nondominant arms for position players (dominant – nondominant)
We assume the players were chosen as random and as the same player is used for both arms, the data is paired. The sample size is small but the boxplot of the differences shows a skewed distribution with no outliers. It is not unreasonable to assume that the data comes from a population that is approximately normal.

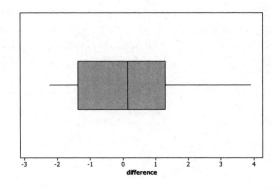

$\bar{d} = 0.2326$, $s_d = 1.6034$, $n = 19$.

The 95% confidence interval for μ_d with $df = 18$ is:

$$\bar{d} \pm (t \text{ critical}) \frac{s_d}{\sqrt{n}} \Rightarrow 0.2326 \pm (2.10) \left(\frac{1.6034}{\sqrt{19}} \right)$$

$$\Rightarrow 0.2326 \pm 0.7872 \Rightarrow (-0.55, 1.02).$$

With 95% confidence, it is estimated that position players using their dominant arm have, an average difference of between 0.55 mm less to 6.1 mm greater, in the translation of their shoulder, than when they use their nondominant arm.

c. No. This question is asking for the difference in the mean translations for the two types of players ie: $\mu_{pitcher} - \mu_{position}$. This is not the same as the mean of the differences: μ_d

11.37 **a.** Let d = the difference in wrist extension while using two different types of computer Mouse. (Type A – Type B)

Let μ_d denote the mean difference for wrist extension between the two types.

H_o: $\mu_d = 0$ H_a: $\mu_d > 0$

= 0.05

Assumptions: Each student used both types of mouse so the data is paired. It is stated that the sample is representative of the population of computer users so it can be seen as a random sample. The sample size is not large ($n = 24$) but without the actual data, we must assume that the population data is approximately normal.

The test statistic is: $t = \dfrac{\bar{x}_d - 0}{\frac{s_d}{\sqrt{n}}}$ with d.f. = 23

$\bar{d} = 8.82$ and $s_d = 10$, $n = 24$

$t = \dfrac{8.82 - 0}{\frac{10}{\sqrt{24}}} = 4.32$

P-value = the area under the 23 df t curve to the right of 4.32   0.00

Since the P-value is less than , the null hypothesis should be rejected at 0.05 There is sufficient evidence to suggest that the mean wrist extension for mouse A is greater than for mouse B.

b. The hypotheses and assumptions will be the same as in Part (a).

The test statistic is: $t = \dfrac{\bar{x}_d - 0}{\frac{s_d}{\sqrt{n}}}$ with d.f. = 23

$\bar{d} = 8.82$ and $s_d = 25$, $n = 24$

$t = \dfrac{8.82 - 0}{\frac{25}{\sqrt{24}}} = 1.73$

P-value = the area under the 23 df t curve to the right of 1.73   0.051.
Although the P-value is just greater than α, the null hypothesis should be not be rejected. However, there is certainly not "convincing evidence" that the mean wrist extension for mouse A is greater than for mouse B.
c. Changing the standard deviation of the differences made the variation in the distribution of the differences greater. So, for the same value of the sample mean difference, the area in the tail was larger and the P-value was greater.

11.39 Let d = the difference in the cost-to-charges ratio for both inpatient and outpatient care.
Let μ_d denote the mean difference in the ratio for 6 hospitals in Oregon in 2002
i.e. Inpatient - Outpatient
H_o: $\mu_d = 0$ H_a: $\mu_d > 0$
 = 0.05
Assumptions: The sample is random and independent. The sample size is small, but a boxplot of the differences shows reasonable symmetry, suggesting that distribution of the population differences are approximately normal.

The test statistic is: $t = \dfrac{\bar{x}_d - 0}{\dfrac{s_d}{\sqrt{n}}}$ with d.f. = 5

$\bar{d}$ = 18.833 and s_d = 5.6716

$t = \dfrac{18.833 - 0}{\dfrac{5.6716}{\sqrt{6}}} = 8.13$

P-value = the area under the 5 df t curve to the right of 8.13)   0.
Thus, the null hypothesis is rejected at 0.05 There is sufficient evidence to suggest that the mean difference cost-to-charge ratio for Oregon hospitals is lower for outpatient care than for inpatient care.

11.41 It is not necessary to use an inference procedure since complete information on all 50 states is available. Inference is necessary only when a sample is selected from some larger population.

11.43 **a.** Let μ_d denote the mean change in blood lactate level for male racquetball players (After Before).
The differences are: 5, 17, 23, 22, 17, 4, 18, 3.
From these, $\bar{x}_d$ = 13.625 and s_d = 8.2797

The 95% confidence interval for $_d$ is

$$\bar{d} \pm (t \text{ critical})\dfrac{s_d}{\sqrt{n}} \quad \Rightarrow \quad 13.625 \pm (2.365)\left(\dfrac{8.2797}{\sqrt{8}}\right)$$

$$\Rightarrow \quad 13.625 \pm 6.923 \quad \Rightarrow \quad (6.702, 20.548).$$

With 95% confidence, it is estimated that the mean change in blood lactate level for male racquetball players is between 6.702 and 20.548.
b. Let μ_d denote the mean change in blood lactate level for female racquetball players (After Before).

The differences are: 10, 10, 6, 3, 0, 20, 7.

From these, $\bar{x}_d = 8.0$ and $s_d = 6.4031$

The 95% confidence interval for μ_d is

$$\bar{d} \pm (t \text{ critical}) \frac{s_d}{\sqrt{n}} \Rightarrow 8.0 \pm (2.45) \left(\frac{6.4031}{\sqrt{7}} \right)$$

$$\Rightarrow 8.0 \pm 5.929 \Rightarrow (2.071, 13.929).$$

With 95% confidence, it is estimated that the mean change in blood lactate level for female racquetball players is between 2.071 and 13.929.

c. Since the two intervals overlap (have values in common), this suggests that it is possible for the mean change for males and the mean change for females to have the same value. (It is appropriate to examine a confidence interval for the difference between the mean changes for men and women to answer this question.)

Exercises 11.44 – 11.59

11.45 Let $_1$ denote the proportion of passengers that flew on airplanes that did not recirculated air that reported post-flight respiratory symptoms, and $_2$ denote the proportion of passengers that flew on airplanes that did recirculated air that reported post-flight respiratory symptoms

$H_o: \pi_1 - \pi_2 = 0 \quad H_a: \pi_1 - \pi_2 \; 0$

$= 0.05$

$$z = \frac{p_1 - p_2}{\sqrt{\frac{p_c(1-p_c)}{n_1} + \frac{p_c(1-p_c)}{n_2}}}$$

$p_1 = \dfrac{108}{517} = 0.2089 \qquad p_2 = \dfrac{111}{583} = 0.1904$

$p_c = \dfrac{n_1 p_1 + n_2 p_2}{n_1 + n_2} = \dfrac{108 + 111}{517 + 583} = 0.1991$

$$z = \frac{(0.2089 - 0.1904)}{\sqrt{\frac{0.1991(1 - 0.1991)}{517} + \frac{0.1991(1 - 0.1991)}{583}}} = \frac{0.0185}{0.0241} = 0.7676$$

P-value = 2(Area under the z curve to the right of 0.77) = 2(0.2206) = 0.4412

Since the P-value is greater than , H_o is not rejected. There is not enough evidence to suggest that the proportion of passengers that reported post-flight respiratory symptoms differs for planes that do and do not recirculate air.

11.47 Let π_1 denote the proportion of Americans aged 12 or older who own an MP3 player in 2006 and π_2 denote the proportion of Americans aged 12 or older who own an MP3 player in 2005.

$n_1 = 1112, \; p_1 = 0.20, \; n_2 = 1112, \; p_2 = 0.15$

The samples are independently selected random samples and both sample sizes are large:

$n_1 p_1 = 1112(.20) = 222.4 \geq 10, \quad n_1(1 - p_1) = 1112(.80) = 889.6 \geq 10$

$n_2 p_2 = 1112(.15) = 166.8 \geq 10, \quad n_2(1 - p_2) = 1112(.85) = 945.2 \geq 10$

The 95% confidence interval for $\pi_1 - \pi_2$ is: $(p_1 - p_2) \pm z_{crit}\sqrt{\dfrac{p_1(1-p_1)}{n_1} + \dfrac{p_2(1-p_2)}{n_2}}$

$(0.20 - 0.15) \pm 1.96\sqrt{\dfrac{0.20(1-0.20)}{1112} + \dfrac{0.15(1-0.15)}{1112}} \Rightarrow 0.05 \pm 1.96(0.0161)$

$\Rightarrow 0.05 \pm 0.0316 \Rightarrow (0.0184, 0.0816)$.

We estimate that the proportion of all Americans the age of 12 or more that owns an MP3 player is between 1.84% and 8.2% higher than the proportion in 2005. We used a method to construct this estimate that captures the true difference in the proportions 95% of the time in repeated sampling. Zero is included in the interval indicating that the proportion of Americans age 12 and older who own an MP3 player was significantly higher in 2006 than in 2005.

11.49 Let π_1 denote the proportion of subjects who experienced gastrointestinal symptoms after eating olestra chips and π_2 denote the proportion of subjects who experienced gastrointestinal symptoms after eating regular chips.

$H_0 : \pi_1 - \pi_2 = 0 \qquad H_a \; \pi_1 - \pi_2 \neq 0$

$= 0.05$

$z = \dfrac{p_1 - p_2}{\sqrt{\dfrac{p_c(1-p_c)}{n_1} + \dfrac{p_c(1-p_c)}{n_2}}}$

$p_1 = \dfrac{90}{563} = 0.160 \quad p_2 = \dfrac{93}{529} = 0.176$

The treatments were assigned at random. Both samples are large:

$n_1 p_1 = 563(.16) = 90 \geq 10, \quad n_1(1-p_1) = 563(.84) = 473 \geq 10$

$n_2 p_2 = 529(.176) = 93 \geq 10, \quad n_2(1-p_2) = 529(.824) = 436 \geq 10$

$p_c = \dfrac{n_1 p_1 + n_2 p_2}{n_1 + n_2} = \dfrac{90 + 93}{563 + 529} = 0.168$

$z = \dfrac{(0.16 - 0.176)}{\sqrt{\dfrac{0.168(0.832)}{563} + \dfrac{0.168(0.832)}{529}}} = \dfrac{-0.016}{.0226} = -0.71$

P-value = 2(area under the z curve to the left of -0.71 = 2(0.2389) = .4778.

Since the P-value is greater than , H_0 should not be rejected. There is no evidence that the proportion of individuals who experience gastrointestinal symptoms after eating olestra chips differs from the proportion who experience symptoms after consuming regular ones.

11.51 **a.** Let π_1 denote the proportion of cardiologists who do not know that carbohydrate was the diet component most likely to raise triglycerides. and π_2 be the corresponding proportion for internists.

$$p_1 = \frac{26}{120} = 0.217, \quad p_2 = \frac{222}{419} = 0.530$$

The samples are independently selected samples and both sample sizes are large:

$$n_1 p_1 = 120(.217) = 26 \geq 10, \quad n_1(1-p_1) = 120(.783) = 94 \geq 10$$

$$n_2 p_2 = 419(.53) = 222 \geq 10, \quad n_2(1-p_2) = 419(.47) = 197 \geq 10$$

Randomness is discussed in part (b)

The 95% confidence interval for $\pi_1 - \pi_2$ is:

$$(p_1 - p_2) \pm z_{crit} \sqrt{\frac{p_1(1-p_1)}{n_1} + \frac{p_2(1-p_2)}{n_2}}$$

$$(0.217 - 0.530) \pm 1.96 \sqrt{\frac{0.217(1-0.217)}{120} + \frac{0.53(1-0.53)}{419}} \quad \Rightarrow \quad -0.313 \pm 1.96(0.0448)$$

$$\Rightarrow \quad -0.313 \pm 0.088 \quad \Rightarrow \quad (-0.401, -0.225).$$

We estimate that that the proportion of all cardiologists who do not know that carbohydrate was the diet component most likely to raise triglycerides is between 22.5% and 40% lower than the corresponding internists. We used a method to construct this estimate that captures the true difference in the proportions 95% of the time in repeated sampling.

b. This is a volunteer sample – the physicians were asked to reply to the questionnaire and 84% of them did not; a very high non-response rate. It may be that of the cardiologists, only those who knew the answers to the questions, sent back the questionnaire and the internists sent them back whether they knew the answers or not.

11.53 Let π_1 denote the proportion of all college graduates than get a sunburn and π_2 be the proportion of all those without a high school degree that than get a sunburn.

$$H_0 : \pi_1 - \pi_2 = 0 \quad H_a \ \pi_1 - \pi_2 > 0$$

$$= 0.05$$

$$z = \frac{p_1 - p_2}{\sqrt{\frac{p_c(1-p_c)}{n_1} + \frac{p_c(1-p_c)}{n_2}}}$$

$p_1 = .43 \ \ n = 200 \qquad p_2 = .25, \ n = 200$

We assume the samples are independently selected random samples and both sample sizes are large:

$$n_1 p_1 = 200(.43) = 86 \geq 10, \quad n_1(1-p_1) = 200(.57) = 114 \geq 10$$

$$n_2 p_2 = 200(.25) = 50 \geq 10, \quad n_2(1-p_2) = 200(.75) = 150 \geq 10$$

$$p_c = \frac{n_1 p_1 + n_2 p_2}{n_1 + n_2} = \frac{86 + 50}{200 + 200} = 0.68$$

$$z = \frac{(0.43 - 0.25)}{\sqrt{\frac{0.68(0.32)}{200} + \frac{0.68(0.32)}{200}}} = \frac{.18}{.0466} = 3.86$$

P-value = area under the z curve to the right of 3.86   0.0001

Since the P-value is less than , H_0 should be rejected. There is convincing evidence that the proportion who experience a sunburn is higher for college graduates than it is for those without a high school degree.

11.55 Let π_1 denote the proportion of exposed dogs that develop lymphoma and π_2 be the corresponding proportion for unexposed dogs.

$p_1 = 0.572$ $n = 827$ $p_2 = 0.146$, $n = 130$

We assume the samples are independently selected random samples and both sample sizes are large:

$n_1 p_1 = 827(.572) = 473.044 \geq 10$, $n_1(1 - p_1) = 827(.428) = 353.956 \geq 10$

$n_2 p_2 = 130(.146) = 18.98 \geq 10$, $n_2(1 - p_2) = 130(.954) = 124.02 \geq 10$

The 95% confidence interval for $\pi_1 - \pi_2$ is: $(p_1 - p_2) \pm z_{crit} \sqrt{\dfrac{p_1(1 - p_1)}{n_1} + \dfrac{p_2(1 - p_2)}{n_2}}$

$(0.572 - 0.146) \pm 1.96 \sqrt{\dfrac{0.572(1 - 0.428)}{827} + \dfrac{0.146(1 - 0.146)}{130}}$ $\Rightarrow$ $.426 \pm 1.96(0.0354)$

$\Rightarrow$ $.426 \pm .069$ $\Rightarrow$ $(.357, .495)$.

We believe that the proportion of exposed dogs that develop lymphoma exceeds that for unexposed dogs by somewhere between .357 and .495. We used a method to construct this estimate that captures the true difference in the proportions 95% of the time in repeated sampling.

11.57 **a.** Let $_1$ denote the proportion of Austrian avid mountain bikers with low sperm counts, and $_2$ denote the proportion of Austrian non-bikers with low sperm counts.

$H_0: \pi_1 - \pi_2 = 0$ $H_a: \pi_1 - \pi_2 > 0$

$= 0.05$ (for demonstration purposes)

$z = \dfrac{p_1 - p_2}{\sqrt{\dfrac{p_c(1 - p_c)}{n_1} + \dfrac{p_c(1 - p_c)}{n_2}}}$

$p_1 = 0.9$ $p_2 = 0.26$

$p_c = \dfrac{n_1 p_1 + n_2 p_2}{n_1 + n_2} = \dfrac{90 + 26}{200} = 0.58$

$z = \dfrac{(0.9 - 0.26)}{\sqrt{\dfrac{0.58(1 - 0.58)}{100} + \dfrac{0.58(1 - 0.58)}{100}}} = \dfrac{0.64}{0.0698} = 9.17$

P-value = Area under the z curve to the right of 9.17   0

Since the P-value is less than , H_0 is rejected. There is enough evidence to suggest that the proportion of Austrian avid mountain bikers with low sperm count is higher than the proportion of Austrian non-bikers.

b. These were not a group of men who where randomly put into one of two treatment groups. It was an observational study and no cause and effect conclusion can be made from such a study.

11.59 The researchers looked at the hypotheses: $H_0: \pi_1 - \pi_2 = 0$ vs. $H_a \; \pi_1 - \pi_2 \; 0$ where $_1$ is the proportion of women who survived after 20 years having had a mastectomy and $_2$ is the proportion of women who survived after 20 years having had a lumpectomy and radiation. If they reported no significant difference, they would have failed to reject the null hypothesis.

Exercises 11.61 – 11.89

11.61 **a.** Let $_1$ be the mean elongation (mm) for the square knot and $_2$ the mean elongation for the Duncan loop when using Maxon thread.

$$H_o: \mu_1 - \mu_2 = 0 \qquad H_a: \mu_1 - \mu_2 \neq 0$$

$= 0.01$ (A value for is not specified in the problem. We use $=0.01$ for illustration.)

Test statistic: $t = \dfrac{(\bar{x}_1 - \bar{x}_2) - 0}{\sqrt{\dfrac{s_1^2}{n_1} + \dfrac{s_2^2}{n_2}}}$

Assumptions: The population distributions are (at least approximately) normal and the two samples are independently selected random samples.

$n_1 = 10$, $\bar{x}_1 = 10.0$, $s_1 = 0.1$, $n_2 = 15$, $\bar{x}_2 = 11.0$, $s_2 = 0.3$

$$t = \frac{(10.0 - 11.0) - 0}{\sqrt{\dfrac{(0.1)^2}{10} + \dfrac{(0.3)^2}{15}}} = \frac{-1.0}{0.083666} = 11.9523$$

$$df = \frac{\left(\dfrac{s_1^2}{n_1} + \dfrac{s_2^2}{n_2}\right)^2}{\dfrac{1}{n_1 - 1}\left(\dfrac{s_1^2}{n_1}\right)^2 + \dfrac{1}{n_2 - 1}\left(\dfrac{s_2^2}{n_2}\right)^2} = \frac{(0.001 + 0.006)^2}{\dfrac{(0.001)^2}{9} + \dfrac{(0.006)^2}{14}} = 18.27$$

So $df = 18$ (rounded down to an integer)

P-value = 2 times the area under the 18 df t curve to the right of 11.9 0.

Since the P-value is less than , the null hypothesis is rejected. The data supports the conclusion that the true mean elongation for the square knot and the Duncan loop differ when using Maxon thread.

b. Let $_1$ be the mean elongation (mm) for the square knot and $_2$ the mean elongation for the for the Duncan loop for the Ticon thread.

$$H_o: \mu_1 - \mu_2 = 0 \qquad H_a: \mu_1 - \mu_2 \neq 0$$

$= 0.01$ (A value for is not specified in the problem. We use $=0.01$ for illustration.)

Test statistic: $t = \dfrac{(\bar{x}_1 - \bar{x}_2) - 0}{\sqrt{\dfrac{s_1^2}{n_1} + \dfrac{s_2^2}{n_2}}}$

Assumptions: The population distributions are (at least approximately) normal and the two samples are independently selected random samples.

$n_1 = 10$, $\bar{x}_1 = 2.5$, $s_1 = 0.06$, $n_2 = 11$, $\bar{x}_2 = 10.9$, $s_2 = 0.4$

$$t = \frac{(2.5 - 10.9) - 0}{\sqrt{\frac{(0.06)^2}{10} + \frac{(0.4)^2}{11}}} = \frac{-8.4}{0.1221} = -68.796$$

$$df = \frac{\left(\frac{s_1^2}{n_1} + \frac{s_2^2}{n_2}\right)^2}{\frac{1}{n_1 - 1}\left(\frac{s_1^2}{n_1}\right)^2 + \frac{1}{n_2 - 1}\left(\frac{s_2^2}{n_2}\right)^2} = \frac{(0.00036 + 0.014545)^2}{\frac{(0.00036)^2}{9} + \frac{(0.014545)^2}{10}} = 10.494$$

So $df = 10$ (rounded down to an integer)

P–value = 2 times the area under the 10 df t curve to the left of –68.8 0.

Since the P-value is less than , the null hypothesis can be rejected. The data supports the conclusion that the true mean elongation for the square knot and the Duncan loop differ when using Ticron thread.

c. Let $_1$ be the mean elongation (mm) for the Maxon thread and $_2$ the mean elongation for the Ticron thread when using the Duncan loop.

$$H_o: \mu_1 - \mu_2 = 0 \qquad H_a: \mu_1 - \mu_2 \neq 0$$

 = 0.01 (A value for   is not specified in the problem. We use =0.01 for illustration.)

Test statistic: $$t = \frac{(\bar{x}_1 - \bar{x}_2) - 0}{\sqrt{\frac{s_1^2}{n_1} + \frac{s_2^2}{n_2}}}$$

Assumptions: The population distributions are (at least approximately) normal and the two samples are independently selected random samples.

$n_1 = 15$, $\bar{x}_1 = 11.0$, $s_1 = 0.3$, $n_2 = 11$, $\bar{x}_2 = 10.9$, $s_2 = 0.4$

$$t = \frac{(11.0 - 10.9) - 0}{\sqrt{\frac{(0.3)^2}{15} + \frac{(0.4)^2}{11}}} = \frac{0.1}{0.143337} = 0.6977$$

$$df = \frac{\left(\frac{s_1^2}{n_1} + \frac{s_2^2}{n_2}\right)^2}{\frac{1}{n_1 - 1}\left(\frac{s_1^2}{n_1}\right)^2 + \frac{1}{n_2 - 1}\left(\frac{s_2^2}{n_2}\right)^2} = \frac{(0.006 + 0.014545)^2}{\frac{(0.006)^2}{14} + \frac{(0.014545)^2}{10}} = 17.790$$

So $df = 17$ (rounded down to an integer)

P-value = 2 times the area under the 17 df t curve to the right of 1.9971 0.247.

Since the P-value is greater than   of 0.01 (or even 0.05), the null hypothesis cannot be rejected. The data do not indicate that there is a difference between the mean elongations for the Maxon thread and the Ticron thread when using the Duncan loop.

11.63 Let μ_1 be the mean relative area of orange for Yarra guppies and μ_2 the mean relative area of orange for Paria guppies.

H_o: $\mu_1 - \mu_2 = 0$ H_a: $\mu_1 - \mu_2 \neq 0$

$\alpha = 0.05$ (A value for α is not specified in the problem. We use $\alpha = 0.05$ for illustration.)

Test statistic: $t = \dfrac{(\bar{x}_1 - \bar{x}_2) - 0}{\sqrt{\dfrac{s_1^2}{n_1} + \dfrac{s_2^2}{n_2}}}$

Assumptions: The population distributions are (at least approximately) normal and the two samples are independently selected random samples.

$n_1 = 30$, $\bar{x}_1 = 0.106$, $s_1 = 0.055$, $n_2 = 30$, $\bar{x}_2 = 0.178$, $s_2 = 0.058$

$$t = \frac{(0.106 - 0.178) - 0}{\sqrt{\dfrac{(0.055)^2}{30} + \dfrac{(0.058)^2}{30}}} = \frac{-0.0720}{0.0145} = -4.9337$$

$$df = \frac{\left(\dfrac{s_1^2}{n_1} + \dfrac{s_2^2}{n_2}\right)^2}{\dfrac{1}{n_1 - 1}\left(\dfrac{s_1^2}{n_1}\right)^2 + \dfrac{1}{n_2 - 1}\left(\dfrac{s_2^2}{n_2}\right)^2} = \frac{(0.000101 + 0.000112)^2}{\dfrac{(0.000101)^2}{29} + \dfrac{(0.000112)^2}{29}} = 57.8$$

So $df = 57$ (rounded down to an integer)

P-value = 2 times the area under the 57 df t curve to the left of -4.9337 $\approx$ 0.0000.

Since the P-value is smaller than α, the null hypothesis is rejected. The data provide strong evidence that there is a difference in the amount of relative area of orange between Yarra guppies and Paria guppies with Yarra guppies having a lower mean relative area of orange.

11.65 **a.** Let μ_1 denote the true mean "campus involvement" score for returning students and μ_2 be the corresponding score for the nonreturning students. Then $\mu_1 - \mu_2$ denotes the difference between the means of campus involvement scores for returning and nonreturning students.

$n_1 = 48$, $\bar{x}_1 = 3.21$, $s_1 = 1.01$, $n_2 = 42$, $\bar{x}_2 = 3.31$, $s_2 = 1.03$

$$V_1 = \frac{s_1^2}{n_1} = \frac{(1.01)^2}{48} = 0.02125 \qquad V_2 = \frac{s_2^2}{n_2} = \frac{(1.03)^2}{42} = 0.02526$$

$$df = \frac{(V_1 + V_2)^2}{\dfrac{V_1^2}{n_1 - 1} + \dfrac{V_2^2}{n_2 - 1}} = \frac{(0.02125 + 0.02526)^2}{\dfrac{(0.02125)^2}{47} + \dfrac{(0.02526)^2}{41}} = 85.94$$

Use df = 85. The *t critical value* is 1.988.

The 95% confidence interval for $\mu_1 - \mu_2$ based on this sample is

$(3.21 - 3.31) \pm 1.988\sqrt{0.02125 + 0.02526} \implies -0.1 \pm 1.988(0.2157) \implies (-0.529, 0.329)$.

Observe that the interval includes 0, and so 0 is one of the plausible values of $\mu_1 - \mu_2$. That is, it is plausible that there is no difference between the mean campus involvement scores for returning and nonreturning students.

b. Let μ_1 be the true mean "personal contact" score for returning students and μ_2 be the corresponding score for the nonreturning students.

$$H_o: \mu_1 - \mu_2 = 0 \quad H_a: \mu_1 - \mu_2 > 0 \text{ (i.e., mean score for nonreturning students is lower)}$$
$$\alpha = 0.01$$

Test statistic: $t = \dfrac{(\bar{x}_1 - \bar{x}_2) - 0}{\sqrt{\dfrac{s_1^2}{n_1} + \dfrac{s_2^2}{n_2}}}$

Assumptions: The sample sizes for the two groups are large (say, greater than 30 for each) and the two samples are independently selected random samples.

$n_1 = 48$, $\bar{x}_1 = 3.22$, $s_1 = 0.93$, $n_1 = 42$, $\bar{x}_2 = 2.41$, $s_2 = 1.03$

$$t = \dfrac{(3.22 - 2.41) - 0}{\sqrt{\dfrac{(0.93)^2}{48} + \dfrac{(1.03)^2}{42}}} = \dfrac{0.81}{0.208} = 3.894$$

$$df = \dfrac{\left(\dfrac{s_1^2}{n_1} + \dfrac{s_2^2}{n_2}\right)^2}{\dfrac{1}{n_1 - 1}\left(\dfrac{s_1^2}{n_1}\right)^2 + \dfrac{1}{n_2 - 1}\left(\dfrac{s_2^2}{n_2}\right)^2} = \dfrac{(0.01802 + 0.02526)^2}{\dfrac{(0.01802)^2}{47} + \dfrac{(0.02526)^2}{41}} = 83.36$$

So $df = 83$ (rounded down to an integer)

P-value = the area under the 83 df t curve to the right of 3.894 $\approx$ 0.0001.

Since the P-value is less than α, the null hypothesis of no difference is rejected. The data provide strong evidence to conclude that the mean "personal contact" score for non-returning students is lower than the corresponding score for the returning students.

11.67 Let μ_1 denote the true mean approval rating for male players and μ_2 the true mean approval rating for female players.

$$H_o: \mu_1 - \mu_2 = 0 \quad H_a: \mu_1 - \mu_2 > 0$$
$$\alpha = 0.01$$

Test statistic: $t = \dfrac{(\bar{x}_1 - \bar{x}_2) - 0}{\sqrt{\dfrac{s_1^2}{n_1} + \dfrac{s_2^2}{n_2}}}$

Assumptions: The sample sizes for each group is large (greater than or equal to 30) and the two samples are independently selected random samples.

$n_1 = 56$, $\bar{x}_1 = 2.76$, $s_1 = 0.44$, $n_2 = 67$, $\bar{x}_2 = 2.02$, $s_2 = 0.41$

$$t = \frac{(2.76 - 2.02) - 0}{\sqrt{\dfrac{(0.44)^2}{56} + \dfrac{(0.41)^2}{67}}} = \frac{0.74}{0.0772} = 9.58$$

$$df = \frac{\left(\dfrac{s_1^2}{n_1} + \dfrac{s_2^2}{n_2}\right)^2}{\dfrac{1}{n_1 - 1}\left(\dfrac{s_1^2}{n_1}\right)^2 + \dfrac{1}{n_2 - 1}\left(\dfrac{s_2^2}{n_2}\right)^2} = \frac{(0.003457 + 0.002509)^2}{\dfrac{(0.003457)^2}{55} + \dfrac{(0.002509)^2}{66}} = 113.8$$

So df = 113 (rounded down to an integer)

P-value = area under the 113 df t curve to the right of 9.58 ≈ 1 − 1 = 0.

Since the P-value is less than , the null hypothesis is rejected. At level of significance 0.05, the data supports the conclusion that the mean approval rating is higher for males than for females.

11.69 **a.** Even though the data are paired the two responses from the same subject, one in 1994 and one in 1995, are most likely not strongly "correlated". The questionnaire asked about alcohol consumption during the "previous week" but the alcohol consumption pattern may vary quite a bit within the same individual from one week to the next which would explain the low correlation between the paired responses.

b. Let μ_d denote the true average difference in the number of drinks consumed by this population between 1994 and 1995 (average for 1994 − average for 1995).

A 95% confidence interval for μ_d is

$$\bar{d} \pm (t \text{ critical})\frac{s_d}{\sqrt{n}} \implies 0.38 \pm (1.985)\left(\frac{5.52}{\sqrt{96}}\right)$$

$$\implies 0.38 \pm 1.1185 \implies (-0.738, 1.498).$$

Since zero is included in the confidence interval, zero is a plausible value for μ_d and hence the data do not provide evidence indicating a decrease in the mean number of drinks consumed.

c. Let μ_d denote the mean difference in number of drinks consumed by non credit card shoppers between 1994 and 1995.

H_o: μ_d = 0 H_a: $\mu_d \neq$ 0

We will compute a P-value for this test.

The test statistic is: $t = \dfrac{\bar{x}_d - 0}{\dfrac{s_d}{\sqrt{n}}}$ with d.f. = 849

$\bar{d}$ = 0.12 and s_d = 4.58

$$t = \frac{0.12 - 0}{\dfrac{4.58}{\sqrt{850}}} = 0.764$$

P-value =2 times (the area under the 849 df t curve to the right of 0.764)   0.445. Thus, the null hypothesis cannot be rejected at any of the commonly used significance levels (e.g., = 0.01, 0.05, or 0.10). There is not sufficient evidence to support the conclusion that the mean number of drinks consumed by the non credit card shoppers has changed between 1994 and 1995.

11.71 Let μ_d denote the true mean weight change.

H_o: $\mu_d = 0$ H_a: $\mu_d > 0$

 = 0.05

The test statistic is: $z = \dfrac{\bar{x}_d - 0}{\dfrac{s_d}{\sqrt{n}}}$

$z = \dfrac{5.15 - 0}{\dfrac{11.45}{\sqrt{322}}} = 8.07$

P-value = area under the z curve to the right of 8.07 = 0.

Since the P-value is less than , H_o is rejected. The data very strongly suggests that the true mean weight change is positive for those who quit smoking.

11.73 Let $_1$ denote the proportion of female Indian False Vampire bats that spend over five minutes in the air before locating food. Let $_2$ denote the proportion of male Indian False Vampire bats that spend over five minutes in the air before locating food.

H_o: $\pi_1 - \pi_2 = 0$ H_a: $\pi_1 - \pi_2 \neq 0$

 = 0.01

$z = \dfrac{p_1 - p_2}{\sqrt{\dfrac{p_c(1-p_c)}{n_1} + \dfrac{p_c(1-p_c)}{n_2}}}$

$p_1 = \dfrac{36}{193} = 0.1865,$ $p_2 = \dfrac{64}{168} = 0.3810,$

$p_c = \dfrac{n_1 p_1 + n_2 p_2}{n_1 + n_2} = \dfrac{36 + 64}{193 + 168} = 0.277$

$z = \dfrac{(0.1865 - 0.3810)}{\sqrt{\dfrac{0.277(0.723)}{193} + \dfrac{0.277(0.723)}{168}}} = \dfrac{-0.1945}{0.0472} = -4.12$

P-value = 2(area under the z curve to the left of -4.12)   2(0) = 0.

Since the P-value is less than , H_o is rejected. There is sufficient evidence in the data to support the conclusion that the proportion of female Indian False Vampire bats who spend over five minutes in the air before locating food differs from that of male Indian False Vampire bats.

11.75 Let π_1 denote the proportion of students in the College of Computing who lose their HOPE scholarship at the end of the first year and let π_2 denote the proportion of students in the Ivan Allen College who lose their HOPE scholarship at the end of the first year.

$H_o: \pi_1 - \pi_2 = 0 \quad H_a: \pi_1 - \pi_2 \neq 0$

We will compute a P-value for this test.

$$z = \frac{p_1 - p_2}{\sqrt{\dfrac{p_c(1-p_c)}{n_1} + \dfrac{p_c(1-p_c)}{n_2}}}$$

$p_1 = 0.532, \quad p_2 = 0.649,$

$$p_c = \frac{n_1 p_1 + n_2 p_2}{n_1 + n_2} = \frac{137(0.532) + 111(0.649)}{137 + 111} = 0.5842$$

$$z = \frac{(0.532 - 0.649)}{\sqrt{\dfrac{0.5842(1-0.5842)}{137} + \dfrac{0.5842(1-0.5842)}{111}}} = \frac{-0.11665}{0.06294} = -1.853$$

P-value = 2 times (the area under the z curve to the left of -1.853)   0.0638.
H_o cannot be rejected at a significance level of 0.05 or smaller. There is not sufficient evidence in the data to support the conclusion that the proportion of students in the College of Computing who lose their HOPE scholarship at the end of one year is different from the proportion for the Ivan Allen College.

11.77 Let π_1 denote the true proportion of returning students who do not take an orientation course and π_2 denote the true proportion of returning students who do take an orientation course.

$n_1 = 94, \quad x_1 = 50, \quad p_1 = \dfrac{50}{94} = 0.5319, \quad n_2 = 94, \quad x_2 = 56, \quad p_2 = \dfrac{56}{94} = 0.5957$

The 95% confidence interval for $\pi_1 - \pi_2$ is

$$(0.5319 - 0.5957) \pm 1.96 \sqrt{\frac{0.5319(0.4681)}{94} + \frac{0.5957(0.4043)}{94}} \Rightarrow -0.0638 \pm 1.96(0.0722)$$

$\Rightarrow -0.0638 \pm 0.1415 \Rightarrow (-0.2053, 0.0777).$

With 95% confidence, it is estimated that the difference between the proportion of returning who do not take an orientation course and the proportion of returning students who do take an orientation course may be as small as 0.2053 to as large as 0.0777.

11.79 Let π_1 denote the proportion of games where a player suffers a sliding injury when stationary bases are used. Let π_2 denote the proportion of games where a player suffers a sliding injury when break-away bases are used.

$H_o: \pi_1 - \pi_2 = 0 \quad H_a: \pi_1 - \pi_2 > 0$

$\quad = 0.01$

$$z = \frac{p_1 - p_2}{\sqrt{\dfrac{p_c(1-p_c)}{n_1} + \dfrac{p_c(1-p_c)}{n_2}}}$$

$$p_1 = \frac{90}{1250} = 0.072, \quad p_2 = \frac{20}{1250} = 0.016,$$

$$p_c = \frac{n_1 p_1 + n_2 p_2}{n_1 + n_2} = \frac{90 + 20}{1250 + 1250} = 0.044$$

$$z = \frac{(0.072 - 0.016)}{\sqrt{\dfrac{0.044(0.956)}{1250} + \dfrac{0.044(0.956)}{1250}}} = \frac{0.056}{0.0082} = 6.83$$

P-value = area under the z curve to the right of 6.83 ≈ 0.

Since the P-value is less than , H_o is rejected. The data suggests that the use of break-away bases reduces the proportion of games in which a player suffers a sliding injury.

11.81 Let μ_1 denote the mean number of goals scored per game for games in which Gretzky played and μ_2 the mean number of goals scored per game for games in which he did not play.

$H_o: \mu_1 - \mu_2 = 0 \quad H_a: \mu_1 - \mu_2 > 0$

$\alpha = 0.01$

Test statistic: $t = \dfrac{(\bar{x}_1 - \bar{x}_2) - 0}{\sqrt{\dfrac{s_1^2}{n_1} + \dfrac{s_2^2}{n_2}}}$

Assumptions: The population distributions are (at least approximately) normal and the two samples are independently selected random samples.

$n_1 = 41, \bar{x}_1 = 4.73, s_1 = 1.29, n_2 = 17, \bar{x}_2 = 3.88, s_2 = 1.18$

$$t = \frac{(4.73 - 3.88) - 0}{\sqrt{\dfrac{(1.29)^2}{41} + \dfrac{(1.18)^2}{17}}} = \frac{0.85}{0.3500} = 2.4286$$

$$df = \frac{\left(\dfrac{s_1^2}{n_1} + \dfrac{s_2^2}{n_2}\right)^2}{\dfrac{1}{n_1 - 1}\left(\dfrac{s_1^2}{n_1}\right)^2 + \dfrac{1}{n_2 - 1}\left(\dfrac{s_2^2}{n_2}\right)^2} = \frac{(0.040588 + 0.081906)^2}{\dfrac{(0.040588)^2}{40} + \dfrac{(0.081906)^2}{16}} = 32.6$$

So $df = 32$ (rounded down to an integer)

P-value = area under the 32 df t curve to the right of 2.4286 ≈ 0.0105.

Since the P-value exceeds , H_o is not rejected. At a significance level of 0.01, the sample data does not support the conclusion that the mean number of goals scored per game is larger when Gretzky played than when he didn't play.

11.83 Let $_1$ denote the mean number of imitations for infants who watch a human model. Let $_2$ denote the mean number of imitations for infants who watch a doll.

$H_o: \mu_1 - \mu_2 = 0 \quad H_a: \mu_1 - \mu_2 > 0$

$\quad = 0.01$

Test statistic: $\quad t = \dfrac{(\bar{x}_1 - \bar{x}_2) - 0}{\sqrt{\dfrac{s_1^2}{n_1} + \dfrac{s_2^2}{n_2}}}$

Assumptions: The population distributions are (at least approximately) normal and the two samples are independently selected random samples.

$n_1 = 12,\ \bar{x}_1 = 5.14,\ s_1 = 1.6,\ n_2 = 15,\ \bar{x}_2 = 3.46,\ s_2 = 1.3$

$t = \dfrac{(5.14 - 3.46) - 0}{\sqrt{\dfrac{(1.6)^2}{12} + \dfrac{(1.3)^2}{15}}} = \dfrac{1.68}{0.570964} = 2.94$

$df = \dfrac{\left(\dfrac{s_1^2}{n_1} + \dfrac{s_2^2}{n_2}\right)^2}{\dfrac{1}{n_1 - 1}\left(\dfrac{s_1^2}{n_1}\right)^2 + \dfrac{1}{n_2 - 1}\left(\dfrac{s_2^2}{n_2}\right)^2} = \dfrac{(0.1024 + 0.0676)^2}{\dfrac{(0.1024)^2}{11} + \dfrac{(0.0676)^2}{14}} = 21.1$

So $df = 21$ (rounded down to an integer)

P-value = area under the 21 df t curve to the right of 2.94   0.0039.

The P-value is less than 0.01 and so the null hypothesis is rejected. The data supports the conclusion that the mean number of imitations by infants who watch a human model is larger than the mean number of imitations by infants who watch a doll.

11.85 Let $_1$ and $_2$ denote the mean score of teenage boys and teenage girls, respectively, on the "worries scale".

$H_o: \mu_1 - \mu_2 = 0 \quad H_a: \mu_1 - \mu_2 > 0$

$\quad = 0.05$

Test statistic: $\quad t = \dfrac{(\bar{x}_1 - \bar{x}_2) - 0}{\sqrt{\dfrac{s_1^2}{n_1} + \dfrac{s_2^2}{n_2}}}$

Assumptions: The sample sizes for each group is large (greater than or equal to 30) and the two samples are independently selected random samples.

$n_1 = 78,\ \bar{x}_1 = 67.59,\ s_1 = 9.7,\ n_2 = 108,\ \bar{x}_2 = 62.05,\ s_2 = 9.5$

$t = \dfrac{(67.59 - 62.05) - 0}{\sqrt{\dfrac{(9.7)^2}{78} + \dfrac{(9.5)^2}{108}}} = \dfrac{5.54}{1.429} = 3.88$

$$df = \frac{\left(\dfrac{s_1^2}{n_1} + \dfrac{s_2^2}{n_2}\right)^2}{\dfrac{1}{n_1-1}\left(\dfrac{s_1^2}{n_1}\right)^2 + \dfrac{1}{n_2-1}\left(\dfrac{s_2^2}{n_2}\right)^2} = \frac{(1.20628 + 0.83565)^2}{\dfrac{(1.20628)^2}{77} + \dfrac{(0.83565)^2}{107}} = 163.999$$

So $df = 163$ (rounded down to an integer)

P-value = area under the 163 df t curve to the right of 3.88 ≈ 0.00008.

Since the P-value is less than , H_o is rejected. It can be concluded that the mean score of teenage boys is larger than the mean score of teenage girls on the "worries scale".

11.87 Let μ_1 denote the true mean self-esteem score for students who are members of Christian groups and μ_2 the true mean self-esteem score for students who are not members of Christian groups.

H_o: $\mu_1 - \mu_2 = 0$ H_a: $\mu_1 - \mu_2 \ne 0$

$\alpha = 0.01$

Test statistic: $t = \dfrac{(\bar{x}_1 - \bar{x}_2) - 0}{\sqrt{\dfrac{s_1^2}{n_1} + \dfrac{s_2^2}{n_2}}}$

Assumptions: The sample sizes for each group is large (greater than or equal to 30) and the two samples are independently selected random samples.

$n_1 = 169$, $\bar{x}_1 = 25.08$, $s_1 = 10$, $n_2 = 124$, $\bar{x}_2 = 24.55$, $s_2 = 8$

$$t = \frac{(25.08 - 24.55) - 0}{\sqrt{\dfrac{(10)^2}{169} + \dfrac{(8)^2}{124}}} = \frac{0.53}{1.052542} = 0.5035$$

$$df = \frac{\left(\dfrac{s_1^2}{n_1} + \dfrac{s_2^2}{n_2}\right)^2}{\dfrac{1}{n_1-1}\left(\dfrac{s_1^2}{n_1}\right)^2 + \dfrac{1}{n_2-1}\left(\dfrac{s_2^2}{n_2}\right)^2} = \frac{(0.59172 + 0.51613)^2}{\dfrac{(0.59172)^2}{168} + \dfrac{(0.51613)^2}{123}} = 288.8$$

So $df = 288$ (rounded down to an integer)

P-value = 2(area under the 288 df t curve to the right of 0.5035) = 2(0.3075) = 0.6150.

Since the P-value exceeds , the null hypothesis is not rejected. The sample data does not support the conclusion that the mean self-esteem score for students who are members of Christian groups differs from that of students who are not members of Christian groups.

11.89 Let π_1 denote the true proportion of adults born deaf who remove the implants. Let π_2 denote the true proportion of adults who went deaf after learning to speak who remove the implants.

$H_o: \pi_1 - \pi_2 = 0$ $H_a: \pi_1 - \pi_2 \neq 0$

$\alpha = 0.01$

$$z = \frac{p_1 - p_2}{\sqrt{\frac{p_c(1-p_c)}{n_1} + \frac{p_c(1-p_c)}{n_2}}}$$

$n_1 = 250$, $x_1 = 75$, $p_1 = 0.3$, $n_2 = 250$, $x_2 = 25$, $p_2 = 0.1$

$$p_c = \frac{n_1 p_1 + n_2 p_2}{n_1 + n_2} = \frac{75 + 25}{250 + 250} = 0.2$$

$$z = \frac{(0.3 - 0.1)}{\sqrt{\frac{0.2(0.8)}{250} + \frac{0.2(0.8)}{250}}} = \frac{0.2}{0.03577} = 5.59$$

P-value = 2(area under the z curve to the right of 5.59) $\approx$ 0.

Since the P-value is less than α, the null hypothesis is rejected. The data does support the fact that the true proportion who remove the implants differs in those that were born deaf from that of those who went deaf after learning to speak.

Chapter 12

Exercises 12.1 – 12.13

12.1 **a.** 0.020 < P-value < 0.025
b. 0.040 < P-value < 0.045
c. 0.035 < P-value < 0.040
d. P-value < 0.001
e. P-value > 0.100

12.3 **a.** df = 3 and χ^2 = 19.0. From Appendix Table 8, P-value < 0.001. Since the P-value is less than α, H_o is rejected.
b. If n = 40, then it is not advisable to use the chi-square test since one of the expected cell frequencies (cell corresponding to nut type 4) would be less than 5.

12.5 **a.** H_0: $\pi_1 = .25$, $\pi_2 = .2$ $\pi_3 = .3$ $\pi_4 = .25$

H_a: H_0 is not true.

α = 0.05 (for demonstration purposes)

Test statistic: $\chi^2 = \sum_{all\ cells} \dfrac{(\text{observed count} - \text{expected count})^2}{\text{expected count}}$

n = 1031
Expected count for each cell = 1031(0.25) = 257.75, 1031(.2) = 206.2, 1031(.3) = 309.3 and 1031(0.25) = 257.75.
The expected counts for each cell are all greater than 5. It is stated that the sample is representative of the male smokers who smoked low tar cigarettes so we can assume that they were randomly sampled.

$$\chi^2 = \frac{(237 - 257.75\,)^2}{257.75} + \frac{(258 - 206.2)^2}{206.2} + \frac{(320 - 309.3)^2}{309.3} + \frac{(216 - 257.75)^2}{257.75}$$
$$= 1.67 + 13.01 + 0.37 + 6.76 = 21.81$$

df = 3. From Appendix Table 8, *P*-value < 0.001. Since the P-value is less than α, H_0 is rejected. There is convincing evidence that the proportion of male low tar cigarette smokers who subsequently died of lung cancer, is not the same for the age category at which they started smoking.
b. If 50% of all the smokers in this population start smoking between the ages of 16 to 20 and if all five years in this age range are equally likely, there should be about 10% starting to smoke for each of the ages 16, 17, 18, 19, 20. This is equivalent to 20% of all the smokers started between the ages of 16 to 17 and 30% of all the smokers started between the ages of 18 to 20.

12.7   **a.** Let π_i denote the true proportion of fatal bicycle accidents in the month they occurred in 2004.

(i = 1, 2, 3 ... 11, 12, where 1 = Jan, 2 = Feb, 3 = March etc.).

$$H_0: \quad \pi_1 = \pi_2 = \pi_3 = \pi_4 = \pi_5 = \pi_6 = \pi_7 = \pi_8 = \frac{1}{12} = 0.083$$

$H_a:$ at least one of the true proportions differs from 0.083.

$\alpha = 0.01$

Test statistic: $\chi^2 = \displaystyle\sum_{\text{all cells}} \dfrac{(\text{observed count} - \text{expected count})^2}{\text{expected count}}$

$n = 719$ Expected count for each cell = 719(0.083) = 59.917.

The expected counts for each cell are all greater than 5. It is stated that the sample is a random sample of fatal bicycle accidents in 2004.

$$\chi^2 = \frac{(38 - 59.917)^2}{59.917} + \frac{(32 - 59.917)^2}{59.917} + \frac{(43 - 59.917)^2}{59.917} + \frac{(59 - 59.917)^2}{59.917} + \frac{(78 - 59.917)^2}{59.917} +$$

$$\frac{(74 - 59.917)^2}{59.917} + \frac{(98 - 59.917)^2}{59.917} + \frac{(85 - 59.917)^2}{59.917} + \frac{(64 - 59.917)^2}{59.917} +$$

$$\frac{(66 - 59.917)^2}{59.917} + \frac{(42 - 59.917)^2}{59.917} + \frac{(40 - 59.917)^2}{59.917}$$

= 8.02 + 13.01 + 4.78 + 0.01 + 5.46 + 3.31 + 24.2 + 10.5 + 0.28 + 5.36 + 6.62 = 81.55

df = 11. From Appendix Table 8, *P*-value < 0.001. Since the P-value is less than α, H_0 is rejected. There is convincing evidence that the proportion of fatal bicycle accidents are not equally likely to occur in each of the 12 months.

b. The proportion for each month would reflect the number of days in the month out of the 366 days in 2004. April, June, September and November have 30 days (30/366 = .082), February has 29 (29/366 = .079) and the rest have 31 (31/366 = .085)

c. Let π_i denote the true proportion of fatal bicycle accidents in the month they occurred in 2004.

(i = 1, 2, 3 ... 11, 12, where 1 = Jan, 2 = Feb, 3 = March etc.).

$H_0: \quad \pi_4 = \pi_6 = \pi_9 = \pi_{11} = .082 \; \pi_2 = .079 \; \pi_1 = \pi_3 = \pi_5 = \pi_7 = \pi_8 = \pi_{10} = \pi_{12} = 0.085$

$H_a:$ at least one of the true proportions differs from H_0.

$\alpha = 0.01$ Test statistic: $\chi^2 = \displaystyle\sum_{\text{all cells}} \dfrac{(\text{observed count} - \text{expected count})^2}{\text{expected count}}$

$n = 719$

Expected count for cells 4, 6, 9 and 11: = 719(0.082) = 58.958, for cell 2: = 719(.079) = 56.801 and for cells 1, 3, 5, 7, 8, 10, and 12: = 719(.085) = 61.115

The expected counts for each cell are all greater than 5. It is stated that the sample is a random sample of fatal bicycle accidents in 2004.

$$\chi^2 = \frac{(38-61.115)^2}{61.115} + \frac{(32-56.801)^2}{56.801} + \frac{(43-61.115)^2}{61.115} + \frac{(59-58.958)^2}{58.958} + \frac{(78-61.115)^2}{61.115} +$$

$$\frac{(74-58.958)^2}{58.958} + \frac{(98-61.115)^2}{61.115} + \frac{(85-61.115)^2}{61.115} + \frac{(64-58.958)^2}{58.958} +$$

$$\frac{(66-61.115)^2}{61.115} + \frac{(42-58.958)^2}{58.958} + \frac{(40-61.115)^2}{61.115}$$

= 8.74 + 10.83 + 5.37 + 0.00003 + 4.67 + 3.84 + 22.26 + 9.33 + 0.43 + 0.39 + 4.88 + 7.30

= 78.04

df = 11. From Appendix Table 8, P-value < 0.001. Since the P-value is less than α, H_0 is rejected. There is convincing evidence that the proportion of fatal bicycle accidents are not equally likely to occur in each of the 12 months taking into account the number of days in the month.

12.9 Let π_i denote the true proportion of birds choosing color i first (i = 1, 2, 3, 4). Here 1=Blue, 2=Green, 3=Yellow, and 4=Red.

H_0: $\pi_1 = \pi_2 = \pi_3 = \pi_4 = 0.25$

H_a: at least one of the true proportions differ from 0.25.

$\alpha = 0.01$

Test statistic: $\chi^2 = \sum_{\text{all cells}} \frac{(\text{observed count} - \text{expected count})^2}{\text{expected count}}$

n = 16 + 8 + 6 + 3 = 33.

Expected count for each cell = 33(0.25) = 8.25.

$$\chi^2 = \frac{(16-8.25)^2}{8.25} + \frac{(8-8.25)^2}{8.25} + \frac{(6-8.25)^2}{8.25} + \frac{(3-8.25)^2}{8.25}$$

= 7.28030 + 0.00758 + 0.61364 + 3.34091 = 11.242

df = 3. From Appendix Table 8, 0.010 < P-value < 0.015. Since the P-value is greater than α, H_0 is not rejected. The data do not provide sufficient evidence indicating a color preference.

12.11 Let π_i denote the proportion of all returned questionnaires accompanied by cover letter i (i = 1, 2, 3).

H_0: $\pi_1 = \frac{1}{3}, \pi_2 = \frac{1}{3}, \pi_3 = \frac{1}{3}$

H_a: H_0 is not true.

$\alpha = 0.05$

Test statistic: $\chi^2 = \sum_{\text{all cells}} \frac{(\text{observed count} - \text{expected count})^2}{\text{expected count}}$

Computations: $n = 131$ $\dfrac{n}{3} = 43.67$

$$\chi^2 = \frac{(48-43.67)^2}{43.67} + \frac{(44-43.67)^2}{43.67} + \frac{(39-43.67)^2}{43.67} = 0.429 + 0.0025 + 0.499 = 0.931$$

df = 2. From Appendix Table 8, P-value > 0.100. Since the P-value exceeds α, the null hypothesis is not rejected. The data does not suggest that the proportions of returned questionnaires differ for the three cover letters.

12.13 Let π_i denote the proportion of phenotype i (i = 1, 2, 3, 4).

H_o: $\pi_1 = \dfrac{9}{16}, \pi_2 = \dfrac{3}{16}, \pi_3 = \dfrac{3}{16}, \pi_4 = \dfrac{1}{16}$

H_a: H_o is not true

$\alpha = 0.01$

Test statistic: $\chi^2 = \displaystyle\sum_{all\ cells} \dfrac{(observed\ count\ -\ expected\ count)^2}{expected\ count}$

Computations:

	Phenotype				
	1	2	3	4	Total
Frequency	926	288	293	104	1611
Expected	906.19	302.06	302.06	100.69	

$$\chi^2 = \frac{(926-906.19)^2}{906.19} + \frac{(288-302.06)^2}{302.06} + \frac{(293-302.06)^2}{302.06}$$

$= 0.433 + 0.655 + 0.278 + 0.109 = 1.47$

df = 3. From Appendix Table 8, P-value > 0.10. Since the P-value exceeds α, the null hypothesis is not rejected. The data appears to be consistent with Mendel's laws.

Exercises 12.15 – 12.35

12.15 a. The d.f. will be (6 - 1)(3 - 1) = 10.
b. The d.f. will be (7 - 1)(3 - 1) = 12.
c. The d.f. will be (6 - 1)(4 - 1) = 15.

12.17 H_0: The proportion falling in the three credit card response categories is the same for all three years. H_a: The proportion falling in the three credit card response categories is not the same for all three years.

$\alpha = 0.05$

Test statistic: $\chi^2 = \displaystyle\sum_{all\ cells} \dfrac{(observed\ count\ -\ expected\ count)^2}{expected\ count}$

Observed and expected frequencies are given in the table below (expected frequencies in parentheses)

Expected cell count = $\dfrac{(row\ total)(column\ total)}{grand\ total}$

	2004	2005	2006	
Definitely/Probably will	40 (43.33)	50 (43.33)	40 (43.33)	130
Might/Might Not/Probably Not	180 (176.67)	190 (176.67)	160 (176.67)	530
Definitely Will Not	781 (781)	761 (781)	801 (781)	2343
	1001	1001	1001	3003

The table contains the expected counts, all of which are greater than 5. Although the data was obtained through a telephone survey, we will assume it consists of independently chosen random samples.

$$\chi^2 = \frac{(40-43.33)^2}{43.33} + \frac{(50-43.33)^2}{43.33} + \frac{(40-43.33)^2}{43.33} + \frac{(180-176.67)^2}{176.67}$$

$$+ \frac{(190-176.67)^2}{176.67} + \frac{(160-176.67)^2}{176.67} + \frac{(781-781)^2}{781} + \frac{(761-781)^2}{781} + \frac{(801-781)^2}{781} = 5.204$$

df = (3 – 1)(3 – 1) = 4. From Appendix Table 8, P-value > 0.1. As the P-value is greater than α, the null hypothesis should not be rejected. There is not enough evidence to suggest that the proportion falling in the three credit card response categories is different for all three years.

12.19 a. H_0: The proportions falling into the each of the hormone use categories is the same for women who have been diagnosed with venous thrombosis and those who have not.

H_a: The proportions falling into the each of the hormone use categories is not the same for women who have been diagnosed with venous thrombosis and those who have not.

α = 0.05 (for demonstration purposes)

Test statistic: $\chi^2 = \sum\limits_{all\ cells} \frac{(observed\ count - expected\ count)^2}{expected\ count}$

Observed and expected frequencies are given in the table below (expected frequencies in parentheses)

$$Expected\ cell\ count = \frac{(row\ total)(column\ total)}{grand\ total}$$

Observed and expected frequencies are given in the table below (expected frequencies in parentheses)

	None	Esterified Estrogen	Conjugated Equine Estrogen	Total
Venous Thrombosis	372 (371.57)	86 (123.31)	121 (84.12)	579
No Venous Thrombosis	1439 (1439.43)	515 (477.69)	289 (325.88)	2243
Total	1811	601	410	2822

The table contains the expected counts, all of which are greater than 5. The data consists of independently chosen random samples from their population.

$$\chi^2 = \frac{(372-371.57)^2}{371.57} + \frac{(86-123.31)^2}{123.31} + \frac{(121-84.12)^2}{84.12} + \frac{(1439-1439.43)^2}{1439.43}$$
$$+ \frac{(515-477.69)^2}{477.69} + \frac{(289-325.88)^2}{325.88} = 34.544$$

df = (3 – 1)(2 – 1) = 2. From Appendix Table 8, P-value < 0.001. Because the P-value is so small, the null hypothesis should be rejected.

It is reasonable to conclude that the proportions falling into the each of the hormone use categories is not the same for women who have been diagnosed with venous thrombosis and those who have not.

b. The results of part (a) could be generalized to the population of all menopausal women who are in the large HMO in the state of Washington.

12.21 H_0: The trust in the President is the same in 2005 as it was in 2002.

H_a: The trust in the President is not the same in 2005 as it was in 2002.

α = 0.05 (for demonstration purposes)

Test statistic: $\chi^2 = \sum\limits_{all\ cells} \dfrac{(observed\ count\ -\ expected\ count)^2}{expected\ count}$

Observed and expected frequencies are given in the table below (expected frequencies in parentheses)

$$Expected\ cell\ count = \frac{(row\ total)(column\ total)}{grand\ total}$$

Response	Year		Total
	2005	2002	
All of the Time	132 (155.48)	180 (156.52)	312
Most of the Time	337 (431.05)	528 (433.95)	865
Some of the Time	554 (473.41)	396 (476.59)	950
Never	169 (132.06)	96 (132.94)	265
Total	1216	1203	2392

The table contains the expected counts, all of which are greater than 5. The data consists of independently chosen random samples of American undergraduates.

$$\chi^2 = \frac{(132-155.48)^2}{155.48} + \frac{(180-156.52)^2}{156.52} + \frac{(337-431.05)^2}{431.05} + \frac{(528-433.95)^2}{433.95} + \frac{(554-473.41)^2}{473.41}$$
$$+ \frac{(396-476.59)^2}{476.59} + \frac{(169-132.06)^2}{132.06} + \frac{(96-132.94)^2}{132.94} = 95.921$$

df = (4 − 1)(2 − 1) = 3. From Appendix Table 8, P-value < 0.001. Because the P-value is so small, the null hypothesis should be rejected. It is reasonable to conclude that the trust in the President is not the same in 2005 as it was in 2002.

12.23 H_0: Region of residence and whether or not the student has a balance exceeding $7000 is independent.

H_a: Region of residence and whether or not the student has a balance exceeding $7000 is not independent.

$\alpha = 0.01$

Test statistic: $\chi^2 = \sum_{all\ cells} \dfrac{(\text{observed count} - \text{expected count})^2}{\text{expected count}}$

Observed and expected frequencies are given in the table below (expected frequencies in parentheses)

Expected cell count = $\dfrac{(\text{row total})(\text{column total})}{\text{grand total}}$

	Balance over $7000?		
Region	**No**	**Yes**	**Total**
Northeast	28 (87.34)	537 (477.66)	565
Midwest	162 (53.18)	182 (290.82)	344
South	42 (80.85)	481 (444.15)	523
West	9 (19.63)	118 (107.37)	127
Total	241	1318	1559

The table contains the expected counts, all of which are greater than 5. The data consists of independently chosen random samples of American undergraduates.

$$\chi^2 = \frac{(28-87.34)^2}{87.34} + \frac{(537-477.66)^2}{477.66} + \frac{(162-53.18)^2}{53.18} + \frac{(182-290.82)^2}{290.82} + \frac{(42-80.85)^2}{80.85}$$

$$+\frac{(481-444.15)^2}{444.15} + \frac{(9-19.63)^2}{19.63} + \frac{(118-107.37)^2}{107.37} = 994$$

df = (4 − 1)(2 − 1) = 3. From Appendix Table 8, P-value < 0.001. Because the P-value is smaller than α, the null hypothesis should be rejected. It is reasonable to conclude that there is an association between region of residence for American undergraduates and whether or not they have a credit card balance of over $7000.

12.25 H_0: Gender and the age at which smokers began smoking are independent.

H_a: Gender and the age at which smokers began smoking are not independent.

$\alpha = 0.05$ (for demonstration purposes)

Test statistic: $\chi^2 = \sum_{all\ cells} \dfrac{(\text{observed count} - \text{expected count})^2}{\text{expected count}}$

Observed and expected frequencies are given in the table below (expected frequencies in parentheses)

Expected cell count = $\dfrac{\text{(row total)(column total)}}{\text{grand total}}$

Age when Smoking began	Gender		Total
	Male	Female	
< 16	25(17.78)	10 (17.22)	35
16-17	24 (20.83)	17 (20.17)	41
18-20	28 (30.48)	32 (29.52)	60
≥ 21	19 (26.92)	34(26.08)	53
Total	96	93	189

The table contains the expected counts, all of which are greater than 5. The data consists of independently chosen random samples of smokers.

$$\chi^2 = \frac{(25-17.78)^2}{17.78} + \frac{(10-17.22)^2}{17.22} + \frac{(24-20.83)^2}{20.83} + \frac{(17-20.17)^2}{20.17} + \frac{(28-30.48)^2}{30.48}$$

$$+ \frac{(32-29.52)^2}{29.52} + \frac{(19-26.92)^2}{26.92} + \frac{(34-26.08)^2}{26.08} = 12.091$$

df = (4 – 1)(2 – 1) = 3. From Appendix Table 8, 0.001 < P-value < 0.005. Because the P-value is smaller than α, the null hypothesis should be rejected. There is sufficient evidence to conclude that there is an association between gender and the age at which a smoker begins smoking.

12.27 H_0: Age group and whether or not the individual is considered in good health are independent .

H_a: Age group and whether or not the individual is considered in good health are not independent.

$\alpha = 0.01$

Test statistic: $\chi^2 = \displaystyle\sum_{\text{all cells}} \dfrac{(\text{observed count } - \text{ expected count})^2}{\text{expected count}}$

Observed and expected frequencies are given in the table below (expected frequencies in parentheses)

Expected cell count = $\dfrac{\text{(row total)(column total)}}{\text{grand total}}$

Age	Health Status		Total
	Good Health	Poor Health	
18-34	920(856.67)	80 (143.33)	1000
35-54	860 (856.67)	140 (143.33)	1000
55-64	790 (856.67)	210 (143.33)	1000
Total	2570	430	3000

The table contains the expected counts, all of which are greater than 5. The data consists of people considered to be representative of American adults.

$$\chi^2 = \frac{(920 - 856.67)^2}{856.67} + \frac{(80 - 143.33)^2}{143.33} + \frac{(860 - 856.67)^2}{856.67} + \frac{(140 - 143.33)^2}{143.33}$$

$$+ \frac{(790 - 856.67)^2}{856.67} + \frac{(210 - 143.33)^2}{143.33} = 68.953$$

$df = (3 - 1)(2 - 1) = 2$. From Appendix Table 8, P-value < 0.001. Because the P-value is smaller than α, the null hypothesis should be rejected. There is sufficient evidence to conclude that there is an association between age group and whether or not the individual is considered in good health.

12.29 **a.** H_o: Gender and workaholism type are independent

H_a: Gender and workaholism type are not independent

$\alpha = 0.05$ (no significance level is given in the problem so we use 0.05 for illustration.)

Test statistic: $\chi^2 = \displaystyle\sum_{all\ cells} \frac{(observed\ count\ -\ expected\ count)^2}{expected\ count}$

Observed and expected frequencies are given in the table below (expected frequencies in parentheses)

		Gender		
		Female	Male	
	Work enthusiasts	20 (27.40)	41 (33.60)	61
Workaholism types	Workaholics	32 (30.99)	37 (38.01)	69
	Enthusiastic workaholics	34 (35.93)	46 (44.07)	80
	Unengaged workers	43 (42.67)	52 (52.33)	95
	Relaxed workers	24 (22.91)	27 (28.09)	51
	Disenchanted workers	37 (30.09)	30 (36.91)	67
		190	233	423

$$\chi^2 = \frac{(20-27.40)^2}{27.40} + \frac{(32-30.99)^2}{30.99} + \frac{(34-35.93)^2}{35.93} + \frac{(43-42.67)^2}{42.67}$$

$$+ \frac{(24-22.91)^2}{22.91} + \frac{(37-30.09)^2}{30.09} + \frac{(41-33.60)^2}{33.60} + \frac{(37-38.01)^2}{38.01}$$

$$+ \frac{(46-44.07)^2}{44.07} + \frac{(52-52.33)^2}{52.33} + \frac{(27-28.09)^2}{28.09} + \frac{(30-36.91)^2}{36.91}$$

$$= 6.852$$

df = $(6-1)(2-1)$ = 5. From Appendix Table 8, P-value > 0.10. Hence the null hypothesis is not rejected. The data are consistent with the hypothesis of no association between gender and workaholism type.

b. Another interpretation of the lack of association between gender and workaholism type is that, for each workaholism category, the true proportion of women who belong to this category is equal to the true proportion of men who belong to this category.

12.31 H_o: There is no dependence (i.e., independent) between handgun purchase within the year prior to death and whether or not the death was a suicide.

H_a: There is a dependence between handgun purchase within the year prior to death and whether or not the death was a suicide.

α = 0.05 (No significance level is given in the problem. We use 0.05 for illustration.)

Test statistic: $\chi^2 = \sum\limits_{all\ cells} \dfrac{(\text{observed count} - \text{expected count})^2}{\text{expected count}}$

Observed and expected frequencies are given in the table below (expected frequencies in parentheses)

	Suicide	Not suicide	
Purchased Handgun	4 (0.27)	12 (15.73)	16
No handgun purchase	63 (66.73)	3921 (3917.27)	3984
	67	3933	4000

$$\chi^2 = \frac{(4-0.27)^2}{0.27} + \frac{(63-66.73)^2}{66.73} + \frac{(12-15.73)^2}{15.73} + \frac{(3921-3917.27)^2}{3917.27} = 53.067$$

df = $(2-1)(2-1)$ = 1. From Appendix Table 8, P-value < 0.001. Hence the null hypothesis is rejected.

The data provide strong evidence to conclude that there is an association between handgun purchase within the year prior to death and whether or not the death was a suicide.

NOTE: One cell has an expected count that is less than 5. The chi-square approximation is probably not satisfactory.

12.33 H_o: The proportion of correct sex identifications is the same for each nose view.

H_a: The proportion of correct sex identifications is not the same for each nose view.

α = 0.05

Test statistic: $\chi^2 = \sum\limits_{\text{all cells}} \dfrac{(\text{observed count} - \text{expected count})^2}{\text{expected count}}$

Nose view

		Front	Profile	Three quarter	
Sex ID	Correct	23 (26)	26 (26)	29 (26)	78
	Not Correct	17 (14)	14 (14)	11 (14)	42
		40	40	40	120

$$\chi^2 = \frac{(23-26)^2}{26} + \frac{(26-26)^2}{26} + \frac{(29-26)^2}{26} + \frac{(17-14)^2}{14} +$$

$$\frac{(14-14)^2}{14} + \frac{(11-14)^2}{14}$$

= 0.346 + 0.000 + 0.346 + 0.643 + 0.000 + 0.643 = 1.978

df = (2 − 1)(3 − 1) = 2. From Appendix Table 8, P-value > 0.10. Since the P-value exceeds α, the null hypothesis is not rejected. The data does not support the hypothesis that the proportions of correct sex identifications differ for the three different nose views.

12.35 H_0:   Job satisfaction and teaching level are independent.

H_a: Job satisfaction and teaching level are dependent.

$\alpha = 0.05$

Test statistic: $\chi^2 = \sum\limits_{\text{all cells}} \dfrac{(\text{observed count} - \text{expected count})^2}{\text{expected count}}$

Computations:

			Job satisfaction		
			Satisfied	Unsatisfied	
		College	74 (63.763)	43 (53.237)	117
Teaching Level		High School	224 (215.270)	171 (179.730)	395
		Elementary	126 (144.967)	140 (121.033)	266
		Total	424	354	778

$$\chi^2 = \frac{(74-62.763)^2}{63.763} + \frac{(43-53.237)^2}{53.237} + \frac{(224-215.270)^2}{215.270}$$

$$+ \frac{(171-179.730)^2}{179.730} + \frac{(126-144.967)^2}{144.967} + \frac{(140-121.023)^2}{121.023}$$

= 1.644 + 1.968 + 0.354 + 0.424 + 2.482 + 2.972 = 9.844

df = $(3-1)(2-1) = 2$. From Appendix Table 8, 0.010 > P-value > 0.005. Since the P-value is less than α, H_o is rejected. The data supports the conclusion that there is a dependence between job satisfaction and teaching level.

Exercises 12.37 – 12.47

12.37 Let π_i denote the true proportion of students graduating from colleges and universities in California in ethnic group i (i = 1 for White, i = 2 for Black, i = 3 for Hispanic, i = 4 for Asian, and i = 5 for other).

H_o: $\pi_1 = 0.507$, $\pi_2 = 0.066$, $\pi_3 = 0.306$, $\pi_4 = 0.108$, $\pi_5 = 0.013$

H_a: at least one of the true proportions differs from the hypothesized value.

$\alpha = 0.01$

Test statistic: $\chi^2 = \sum\limits_{all\ cells} \dfrac{(\text{observed count } - \text{ expected count})^2}{\text{expected count}}$

n = 1000

The expected cell counts are calculated as follows:

$\left(\begin{array}{c}\text{Expected number of graduates}\\ \text{for ethnic group i}\end{array}\right) = 1000 \times \left(\begin{array}{c}\text{proportion of graduates from ethnic group i}\\ \text{as given in the census report}\end{array}\right)$

Ethnic group	Number of graduates in the sample	Population proportion according to census report	Expected number of graduates in the sample
White	679	0.507	507
Black	51	0.066	66
Hispanic	77	0.306	306
Asian	190	0.108	108
Other	3	0.013	13

$\chi^2 = \dfrac{(679-507)^2}{507} + \dfrac{(51-66)^2}{66} + \dfrac{(77-306)^2}{306} + \dfrac{(190-108)^2}{108} + \dfrac{(3-13)^2}{13}$

= 58.351 + 3.409 + 171.376 + 62.259 + 7.692 = 303.09

df =4. From Appendix Table 8, P-value < 0.001. At a significance level of $\alpha = 0.01$, H_o is rejected. The data provide very strong evidence to conclude that the proportions of students graduating from colleges and universities in California differ from the respective proportions in the population.

12.39 **a.** The table below gives the row percentages for each smoking category.

	< 1/wk	1/wk	2-4/wk	5-6/wk	1/day
Never smoked	33.00	15.79	22.42	11.17	17.62
Smoked in the past	18.53	12.41	23.22	14.68	31.17
Currently smokes	21.58	11.90	19.36	12.19	34.97

The proportions falling into each category appear to be dissimilar. For instance, only 17.62% of the subjects in the "never smoked" category consumed one drink per day, whereas 34.97% of those in the "currently smokes" category consume one drink per day. Similar discrepancies are seen for other categories as well.

b. H_o: Smoking status and alcohol consumption are independent.

H_a: Smoking status and alcohol consumption are not independent.

$\alpha = 0.05$ (No significance level is not given in the problem. We use 0.05 for illustration.)

Test statistic: $\chi^2 = \sum\limits_{all\ cells} \dfrac{(observed\ count\ -\ expected\ count)^2}{expected\ count}$

Observed and expected frequencies are given in the table below (expected frequencies in parentheses)

		Alcohol consumption (no. of drinks)					
		<1/wk	1/wk	2-4/wk	5-6/wk	1/day	
	Never Smoked	3577 (2822.22)	1711 (1520.12)	2430 (2427.33)	1211 (1372.96)	1910 (2696.37)	10839
Smoking status	Smoked in the past	1595 (2241.58)	1068 (1207.37)	1999 (1927.93)	1264 (1090.49)	2683 (2141.62)	8609
	Currently Smokes	524 (632.19)	289 (340.51)	470 (543.74)	296 (307.55)	849 (604.00)	2428
		5696	3068	4899	2771	5442	21876

$$\chi^2 = \frac{(3577-2822.22)^2}{2822.22} + \frac{(1711-1520.12)^2}{1520.12} + \frac{(2430-2427.33)^2}{2427.33} + \frac{(1211-1372.96)^2}{1372.96}$$

$$+\frac{(1910-2696.37)^2}{2696.37} + \frac{(1595-2241.58)^2}{2241.58} + \frac{(1068-1207.37)^2}{1207.37} + \frac{(1999-1927.93)^2}{1927.93}$$

$$+\frac{(1264-1090.49)^2}{1090.49} + \frac{(2683-2141.62)^2}{2141.62} + \frac{(524-632.19)^2}{632.19} + \frac{(289-340.51)^2}{340.51}$$

$$+\frac{(470-543.74)^2}{543.74} + \frac{(296-307.55)^2}{307.55} + \frac{(849-604.00)^2}{604.00} = 980.068$$

df = (3 – 1)(5 – 1) = 8. From Appendix Table 8, P-value < 0.001. Hence the null hypothesis is rejected. The data provide strong evidence to conclude that smoking status and alcohol consumption are not independent.

c. The result of the test in part **b** is consistent with our observations in part **a**.

185

12.41 H_o: Position and Role are independent.

H_a: Position and Role are not independent.

$\alpha = 0.01$

Test statistic: $\chi^2 = \sum\limits_{all\ cells} \dfrac{(\text{observed count} - \text{expected count})^2}{\text{expected count}}$

Observed and expected frequencies are given in the table below (expected frequencies in parentheses)

	Initiate chase	Participate in chase	
Center position	28 (39.04)	48 (36.96)	76
Wing position	66 (54.96)	41 (52.04)	107
	94	89	183

$$\chi^2 = \frac{(28-39.04)^2}{39.04} + \frac{(66-54.96)^2}{54.96} + \frac{(48-36.96)^2}{36.96} + \frac{(41-52.04)^2}{52.04} = 10.97$$

df = (2 – 1)(2 – 1) = 1. From Appendix Table 8, P-value < 0.001. Hence the null hypothesis is rejected. The data provide strong evidence to conclude that there is an association between position and role.

For the chi-square analysis to be valid, the observations on the 183 lionesses in the sample are assumed to be independent.

12.43 H_o: There is no dependence between response and region of residence.

H_a: There is a dependence between response and region of residence.

$\alpha = 0.01$

Test statistic: $\chi^2 = \sum\limits_{all\ cells} \dfrac{(\text{observed count} - \text{expected count})^2}{\text{expected count}}$

		Response		
		Agree	Disagree	
	Northeast	130 (150.35)	59 (38.65)	189
Region	West	146 (149.55)	42 (38.45)	188
	Midwest	211 (209.22)	52 (53.78)	263
	South	291 (268.88)	47 (69.12)	338
		778	200	978

$$\chi^2 = \frac{(130-150.35)^2}{150.35} + \frac{(59-38.65)^2}{38.65} + \frac{(146-149.55)^2}{149.55} + \frac{(42-38.45)^2}{38.45} +$$

$$\frac{(211-209.22)^2}{209.22} + \frac{(52-53.78)^2}{53.78} + \frac{(291-268.88)^2}{268.88} + \frac{(47-69.12)^2}{69.12}$$

= 2.754 + 10.714 + 0.084 + 0.329 + 0.015 + 0.059 + 1.820 + 7.079 = 22.855

df = (4 − 1)(2 − 1) = 3. From Appendix Table 8, 0.001 > P-value.

Since the P-value is less than α, the null hypothesis is rejected.

The data supports the conclusion that there is a dependence between response and region of residence.

12.45   **a.** It is a test of homogeneity since the number of males as well as the number of females were fixed prior to sampling.

     b. H_o: Proportions of each type of offense is the same for males and females.

         H_a: H_o is not true.

         $\alpha = 0.05$

Test statistic: $\chi^2 = \displaystyle\sum_{all\ cells} \frac{(\text{observed count} - \text{expected count})^2}{\text{expected count}}$

Type of crime		Violent	Property	Drug	Public order	Total
	Male	117 (91.5)	150 (155)	109 (138.5)	124 (115)	500
Sex						
	Female	66 (91.5)	160 (155)	168 (138.5)	106 (115)	500
	Total	183	310	277	230	1000

$$\chi^2 = \frac{(117-91.5)^2}{91.5} + \frac{(150-155)^2}{155} + \frac{(109-138.5)^2}{138.5}$$

$$+ \frac{(124-115)^2}{115} + \frac{(66-91.5)^2}{91.5} + \frac{(160-155)^2}{155}$$

= 28.51

df = (4 − 1)(2 − 1) = 3. From Appendix Table 8, 0.001 > P-value. Since the P-value is less than α, the null hypothesis is rejected. The proportions of crimes in each category are not the same for males and females.

12.47 H_o: The ability of individuals to make correct identifications does not differ for the brands of cola.

     H_a: The ability of individuals to make correct identifications differs for the brands of cola.

     $\alpha = 0.05$

Test statistic: $\chi^2 = \displaystyle\sum_{all\ cells} \frac{(\text{observed count} - \text{expected count})^2}{\text{expected count}}$

Computations:

	Number of correct identifications				
	0	1	2	3 or 4	Total
Coca-Cola	13 (14.3)	23 (23.7)	24 (23)	19 (18)	79
Pepsi-Cola	12 (14.3)	20 (23.7)	26 (23)	21 (18)	79
Royal Crown	18 (14.3)	28 (23.7)	19 (23)	14 (18)	79
Total	43	71	69	54	237

$$\chi^2 = \frac{(13-14.3)^2}{14.3} + \frac{(23-23.7)^2}{23.7} + \frac{(24-23)^2}{23} + \frac{(19-18)^2}{18} +$$

$$\frac{(12-14.3)^2}{14.3} + \frac{(20-23.7)^2}{23.7} + \frac{(26-23)^2}{23} + \frac{(21-18)^2}{18} +$$

$$\frac{(18-14.3)^2}{14.3} + \frac{(28-23.7)^2}{23.7} + \frac{(19-23)^2}{23} + \frac{(14-18)^2}{18}$$

= 0.12 + 0.02 + 0.04 + 0.06 + 0.38 + 0.57 + 0.39 + 0.50 + 0.94 + 0.79 + 0.70 + 0.89 = 5.4

df = (3 – 1)(4 – 1) = 6. From Appendix Table 8, P-value > 0.10. Since the P-value exceeds α, the null hypothesis is not rejected. The data are consistent with the hypothesis that the ability of individuals to make correct identifications does not differ for the three brands of cola.

Chapter 13

Exercises 13.1 – 13.11

13.1 **a.** y = 5.0 + 0.017x

      b. When x = 1000, y = 5 + 0.017(1000) = 12

      When x = 2000, y = 5 + 0.017(2000) = 29

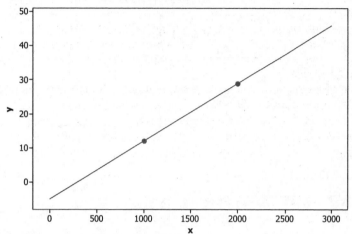

      c. When x = 2100, y = 5 + (0.017)(2100) = 30.7

      d. 0.017

      e. 0.017(100) = 1.7

      f. It is stated that the community where the given regression model is valid has no small houses. Therefore, there is no assurance that the model is adequate for predicting usage based on house size for small houses. Consequently, it is not advisable to use the model to predict usage for a 500 sq.ft house.

13.3 **a.** The mean value of serum manganese when Mn intake is 4.0 is 2 + 1.4(4) = 3.6.

      The mean value of serum manganese when Mn intake is 4.5 is 2 + 1.4(4.5) = 4.3.

      b. $\dfrac{5-3.6}{1.2} = 1.17$

      P(serum Mn over 5) = P(1.17 < z) = 1 − 0.8790 = 0.121.

      c. The mean value of serum manganese when Mn intake is 5 is 2 + 1.4(5) = 5.

         $\dfrac{5-5}{1.2} = 0,$ $\dfrac{3.8-5}{1.2} = -1$

         P(serum Mn over 5) = P(0 < z) = 0.5

         P(serum Mn below 3.8) = P(z < −1) = 0.1587

13.5 **a.** The expected change in price associated with one extra square foot of space is 47. The expected change in price associated with 100 extra square feet of space is 47(100) = 4700.

      b. When x = 1800, the mean value of y is 23000 + 47(1800) = 107600.

$$\frac{110000-107600}{5000}=0.48 \qquad \frac{100000-107600}{5000}=-1.52$$

P(y > 110000) = P(0.48 < z) = 1   0.6844 = 0.3156
P(y < 100000) = P(z <   1.52) = 0.0643
Approximately 31.56% of homes with 1800 square feet would be priced over 110,000 dollars and about 6.43% would be priced under 100,000 dollars.

13.7   **a.** $r^2 = 1 - \frac{\text{SSResid}}{\text{SSTo}} = 1 - \frac{27.890}{73.937} = 1 - 0.3772 = 0.6228$

b. $s_e^2 = \frac{\text{SSResid}}{n-2} = 1 - \frac{27.890}{13-2} = \frac{27.890}{11} = 2.5355$

$s_e = \sqrt{2.5355} = 1.5923$

The magnitude of a typical deviation of residence half-time (y) from the population regression line is estimated to be about 1.59 hours.

c. b = 3.4307

d. $\hat{y}$ = 0.0119 + 3.4307(1) = 3.4426

13.9   **a.** $r^2 = 1 - \frac{2620.57}{22398.05} = 0.883$

b. $s_e = \sqrt{\frac{2620.57}{14}} = \sqrt{187.184} = 13.682$ with 14 df.

13.11   **a.**

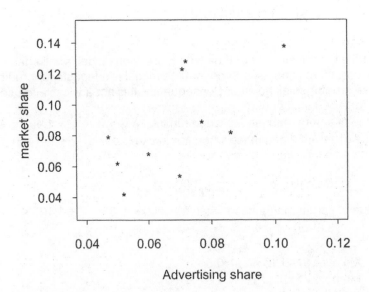

There seems to be a general tendency for y to increase at a constant rate as x increases. However, there is also quite a bit of variability in the y values.

b. Summary values are: n = 10, x = 0.688, x^2 = 0.050072, y = 0.835,
y^2 = 0.079491, xy = 0.060861.

$$b = \frac{0.060861 - \left[\dfrac{(0.688)(0.835)}{10}\right]}{0.050072 - \left[\dfrac{(0.688)^2}{10}\right]} = \frac{0.003413}{0.0027376} = 1.246712$$

a = 0.0835 (1.246712)(0.0688) = 0.002274
The equation of the estimated regression line is ŷ = 0.002274 + 1.246712x.
The predicted market share, when advertising share is 0.09, would be
0.002274 + 1.246712(0.09) = 0.10993.

c. $SSTo = 0.079491 - \left[\dfrac{(0.835)^2}{10}\right] = 0.0097685$

SSResid = 0.079491 (0.002274)(0.835) (1.246712)(0.060861) = 0.0055135
$r^2 = 1 - \dfrac{0.0055135}{0.0097685} = 1 - 0.564 = 0.436$

This means that 43.6% of the total variability in market share (y) can be explained by the
simple linear regression model relating market share and advertising share (x).

d. $s_e = \sqrt{\dfrac{0.0055135}{8}} = \sqrt{0.000689} = 0.0263$ with 8 df.

Exercises 13.13 – 13.25

13.13 **a.** $\sum(x - \bar{x})^2 = 250$ $\sigma_b = \dfrac{4}{\sqrt{250}} = 0.253$

b. $\sum(x - \bar{x})^2 = 500$ $\sigma_b = \dfrac{4}{\sqrt{500}} = 0.179$

No, the resulting value of $_b$ is not half of what it was in **a**. However, the resulting value
of σ_b^2 is half of what it was in **a**.

c. It would require 4 observations at each x value to yield a value of $_b$ which is half the value
calculated in **a**. In this case $\sum(x - \bar{x})^2 = 1000$, so $\sigma_b = \dfrac{4}{\sqrt{1000}} = 0.1265$

13.15 **a.** $s_e = \sqrt{\dfrac{1235.47}{13}} = \sqrt{95.036} = 9.7486$

$s_b = \dfrac{9.7486}{\sqrt{4024.2}} = \dfrac{9.7486}{63.4366} = 0.1537$

b. The 95% confidence interval for is 2.5 (2.16)(0.1537) 2.5 0.33 (2.17, 2.83).
c. The interval is relatively narrow. However, whether   has been precisely estimated or not
depends on the particular application we have in mind.

13.17 **a.** $S_{xy} = 44194 - \left[\dfrac{(50)(16705)}{20}\right] = 2431.5$

$S_{xx} = 150 - \left[\dfrac{(50)^2}{20}\right] = 25$

$b = \dfrac{2431.5}{25} = 97.26, \quad a = 835.25 \quad (97.26)(2.5) = 592.1$

b. $\hat{y} = 592.1 + 97.26(2) = 786.62$. The corresponding residual is
$(y \quad \hat{y}) = 757 \quad 786.62 = \ 29.62$.

c. SSResid $= 14194231 \quad 592.1(16705) \quad 97.26(44194) = 4892.06$

$s_e = \sqrt{\dfrac{4892.06}{18}} = \sqrt{271.781} = 16.4858$

$s_b = \dfrac{16.4858}{\sqrt{25}} = 3.2972$

The 99% confidence interval for , the true average change in oxygen usage associated with a one-minute increase in exercise time is
97.26 (2.88)(3.2972) 97.26 9.50 (87.76, 106.76).

13.19 **a.** H_o: $= 0$ H_a: 0
$= 0.05$ (for illustration)

$t = \dfrac{b}{s_b}$ with df = 42

$t = \dfrac{15}{5.3} = 2.8302$

P-value = 2(area under the 42 df t curve to the right of 2.83) 2(0.0036) = 0.0072.
Since the P-value is less than , H_o is rejected. The data supports the conclusion that the simple linear regression model specifies a useful relationship between x and y. (It is advisable to examine a scatter plot of y versus x to confirm the appropriateness of a straight line model for these data).

b. b (t critical) s_b 15 (2.02)(5.3) 15 10.706 (4.294, 25.706)
Based on this interval, we estimate the change in mean average SAT score associated with an increase of $1000 in expenditures per child is between 4.294 and 25.706.

13.21 Summary values are: n = 10,   x = 6,970 x^2 = 5,693,950   y = 10,148 y^2 =
12,446,748   xy = 8,406,060.
We first calculate various quantities needed to answer the different parts of this problem.

$$b = \cfrac{8406060 - \left[\cfrac{(6970)(10148)}{10}\right]}{5693950 - \left[\cfrac{(6970)^2}{10}\right]} = \frac{1332904}{835860} = 1.5946498217404828560$$

$$a = \frac{[1014.8 - (1.5946498217404828560)(697)]}{10} = -96.670925753116550619$$

SSResid = 12446748   (96.670925753116550619)(10148)   (1.5946498217404828560) (8406060) = 23042.474

NOTE: Using the formula $SSResid = \sum y^2 - a \sum y - b \sum xy$ can lead to severe roundoff errors unless many significant digits are carried along for the intermediate calculations. The calculations for this problem are particularly prone to roundoff errors because of the large numbers involved. This is the reason we have given many significant digits for the slope and the intercept estimates. You may want to try doing these calculations with fewer significant digits. You will notice a substantial loss in accuracy in the final answer. The formula $SSResid = \sum (y - \hat{y})^2$ provides a more numerically stable alternative. We give the calculations based on this alternative formula in the following table :

We used $b = 1.59465$ and $a = \frac{[1014.8 - (1.59465)(697)]}{10} = -96.6711$ to calculate

$\hat{y} = a + bx$.

y	$\hat{y} = a + bx$	$y - \hat{y}$	$(y - \hat{y})^2$
303	301.99	1.009	1
491	477.4	13.597	184.9
659	660.79	-1.788	3.2
683	740.52	-57.52	3308.6
922	876.07	45.935	2110
1044	1083.37	-39.37	1550
1421	1306.62	114.379	13082.6
1329	1370.41	-41.407	1714.5
1481	1513.93	-32.925	1084.1

The sum of the numbers in the last column gives $SSResid$ = 23042.5 which is accurate to the first decimal place.

$$s_e^2 = \frac{23042.5}{8} = 2880.31$$

$$s_b^2 = \frac{2880.31}{835860} = 0.00344592, \quad s_b = 0.0587020$$

a. The prediction equation is CHI = -96.6711 + 1.59465 Control .
Using this equation we can predict the mean response time for those suffering a closed-head injury using the mean response time on the same task for individuals with no head injury.

b. Let β denote the expected increase in mean response time for those suffering a closed-head injury associated with a one unit increase in mean response time for the same task for individuals with no head injury.

$\qquad$ H_o: $\beta = 0$ $\qquad$ H_a: $\beta \neq 0$

$\qquad$ $\alpha = 0.05$

$\qquad$ $t = \dfrac{b}{s_b}$ $\qquad$ with df = 8

$\qquad$ $t = \dfrac{1.59465}{0.0587020} = 27.1652$

$\qquad$ P-value = 2(area under the 8 df t curve to the right of 27.1652) $\approx$ 2(0) = 0.
Since the P-value is less than α, H_o is rejected. The simple linear regression model does provide useful information for predicting mean response times for individuals with CHI and mean response times for the same task for individuals with no head injury. (A scatter plot of y versus x confirms that a straight line model is a reasonable model for this problem).

c. The equation CHI = 1.48 Control says that the mean response time for individuals with CHI is *proportional* to the mean response time for the same task for individuals with no head injury, and the proportionality constant is 1.48. This implies that the mean response time for individuals with CHI is *1.48 times* the mean response time for the same task for individuals with no head injury.

13.23 **a.** Let β denote the expected change in sales revenue associated with a one unit increase in advertising expenditure.

$\qquad$ H_o: $\beta = 0$ $\qquad$ H_a: $\beta \neq 0$

$\qquad$ $\alpha = 0.05$

$\qquad$ $t = \dfrac{b}{s_b}$ $\qquad$ with df = 13 $\qquad$ $t = \dfrac{52.57}{8.05} = 6.53$

P-value = 2(area under the 13 df t curve to the right of 6.53) $\approx$ 2(0) = 0.
Since the P-value is less than α, H_o is rejected. The simple linear regression model does provide useful information for predicting sales revenue from advertising expenditures.

b. H_o: $\beta = 40$ $\qquad$ H_a: $\beta > 40$

$\qquad$ $\alpha = 0.01$

$\qquad$ Test statistic: $t = \dfrac{b - 40}{s_b}$ $\qquad$ with df = 13

$\qquad$ $t = \dfrac{(52.57 - 40)}{8.05} = 1.56$

$\qquad$ P-value = area under the 13 df t curve to the right of 1.56 $\approx$ 0.071.
Since the P-value exceeds α, the null hypothesis is not rejected. The data are consistent with the hypothesis that the change in sales revenue associated with a one unit increase in advertising expenditure does not exceed 40 thousand dollars.

13.25 Let β denote the average change in milk pH associated with a one unit increase in temperature.

H_o: $\beta = 0$ H_a: $\beta < 0$

$\alpha = 0.01$

The test statistic is: $t = \dfrac{b}{s_b}$ with d.f. = 14.

Computations: $n = 16$, $\sum x = 678$, $\sum y = 104.54$,

$$S_{xy} = 4376.36 - \frac{(678)(104.54)}{16} = -53.5225$$

$$S_{xx} = 36056 - \frac{(678)^2}{16} = 7325.75$$

$$b = \frac{-53.5225}{7325.75} = -0.0073$$

$a = 6.53375 - (-0.0073)(42.375) = 6.8431$

SSResid = $683.447 - 6.8431(104.54) - (-0.00730608)(4376.36) = 0.0177354$

$$s_e = \sqrt{\frac{.0177354}{14}} = \sqrt{.001267} = .0356$$

$$s_b = \frac{.0356}{\sqrt{7325.75}} = .000416$$

$$t = \frac{-0.00730608}{0.000416} = -17.5627$$

P-value = area under the 14 df t curve to the left of $-17.5627 \approx 0$.

Since the P-value is less than α, H_0 is rejected. There is sufficient evidence in the sample to conclude that there is a negative (inverse) linear relationship between temperature and pH. A scatter plot of y versus x confirms this.

13.27

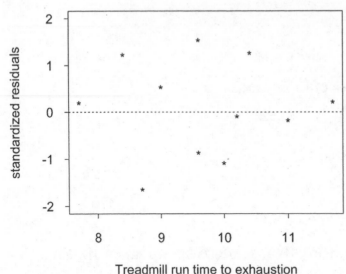

Treadmill run time to exhaustion

The standardized residual plot does not exhibit any unusual features.

13.29 **a.** The assumptions required in order that the simple linear regression model be appropriate are

      (i)   The distribution of the random deviation e at any particular x value has mean value 0.

      (ii)   The standard deviation of e is the same for any particular value of x.

      (iii)   The distribution of e at any particular x value is normal.

      (iv)   The mean value of vigor is a linear function of stem density.

      (v)   The random deviations $e_1, e_2, \cdots, e_n$ associated with different observations are independent of one another.

b.

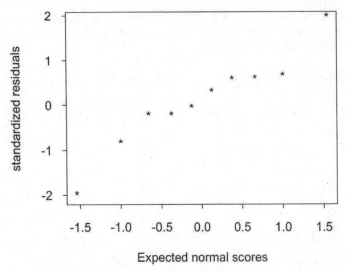

The normal probability plot appears to follow a straight line pattern (approximately). Hence the assumption that the random deviation distribution is normal is plausible.

c.

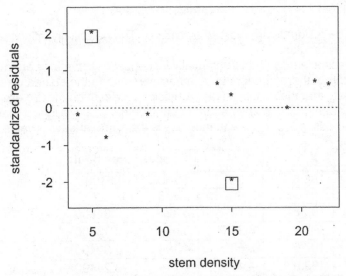

There are two residuals that are relatively large. The corresponding points are enclosed in boxes on the graph above.

d. The negative residuals appear to be associated with small x values, and the positive residuals appear to be associated with large *x* values. Such a pattern is apparently the result of the fitted regression line being influenced by the two potential outlying points. This would cause one to question the appropriateness of using a simple linear regression model without addressing the issue of outliers.

13.31 a.

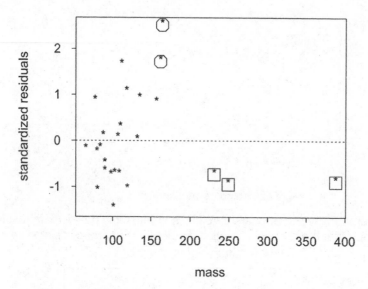

The several large residuals are marked by circles. The potentially influential observations are marked by rectangles.

b. The residuals associated with the potentially influential observations are all negative. Without these three, there appears to be a positive trend to the standardized residual plot. The plot suggests that the simple linear regression model might not be appropriate.

c. There does not appear to be any pattern in the plot that would suggest that it is unreasonable to assume that the variance of *y* is the same at each *x* value.

13.33

Year	X	Y	Y-Pred	Residual
1963	188.5	2.26	1.750	0.51000
1964	191.3	2.60	2.478	0.12200
1965	193.8	2.78	3.128	0.34800
1966	195.9	3.24	3.674	0.43400
1967	197.9	3.80	4.194	0.39400
1968	199.9	4.47	4.714	0.24400
1969	201.9	4.99	5.234	0.24400
1970	203.2	5.57	5.572	0.00200
1971	206.3	6.00	6.378	0.37800
1972	208.2	5.89	6.872	0.98200
1973	209.9	8.64	7.314	1.32600

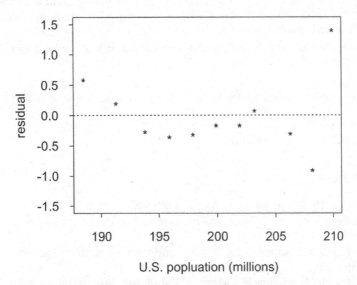

residual

U.S. popluation (millions)

The residuals are positive in 1963 and 1964, then they are negative from 1965 through 1972, followed by a positive residual in 1973. The residuals exhibit a pattern in the plot and thus the plot casts doubt on the appropriateness of the simple linear regression model.

Exercises 13.35 – 13.47

13.35 If the request is for a confidence interval for   the wording would likely be "estimate the <u>change</u> in the average y value associated with a one unit increase in the x variable." If the request is for a confidence interval for   + x* the wording would likely be "estimate the <u>average y</u> value when the value of the x variable is x*."

13.37 $a + b(.6) = -0.6678 + 1.3957(.6) = 0.16962$

$$s_{a+b(.6)} = 0.8123\sqrt{\frac{1}{12} + \frac{(.6 - .485)^2}{.3577}} = 0.2817$$

The 95% confidence interval for   + (.6) is 0.16962   2.23(0.2817)
          0.16962   0.62819   (-.45857, .79781).

13.39 a. The 95% prediction interval for an observation to be made when x* = 40 would be

$$6.5511 \quad 2.15\sqrt{(0.0356)^2 + (0.008955)^2} \quad 6.5511 \quad 2.15(0.0367)$$

          6.5511   0.0789 = (6.4722, 6.6300).

b. The 99% prediction interval for an observation to be made when x* = 35 would be

$$6.5876 \quad 2.98\sqrt{(0.0356)^2 + (0.009414)^2} \quad 6.5876 \quad 2.98(0.0368)$$

          6.5876   0.1097 = (6.4779, 6.6973).

c. Yes, because x* = 60 is farther from the mean value of x, which is 42.375, than is 40 or 35.

13.41 **a.** From the MINITAB output we get
Clutch Size $\tilde{=}$ 133.02 + 5.92 Snout-Vent Length

b. From the MINITAB output we get s_b =1.127

c. Let denote the mean increase in Clutch size associated with a one unit increase in Snout-Vent length.

H_o: = 0 H_a: > 0

= 0.05 (a significance level is not specified in the problem so we use 0.05 for illustration)

The test statistic is: $t = \dfrac{b}{s_b}$ with df. = 12.

From the MINITAB output $t = \dfrac{5.919}{1.127} = 5.25$

P-value = area under the 12 df t curve to the right of 5.25 $\approx$ 0.
Hence the null hypothesis is rejected. The data provide strong evidence indicating that the slope is positive.

d. The predicted value of the clutch size for a salamander with snout-vent length of 65 is – 133.02 + 5.919 (65) = 251.715.

e. The value 105 is very much outside the range of snout-vent length values in the available data. The validity of the estimated regression line this far away from the range of x values in the data set is highly questionable. Therefore, calculation of a predicted value and/or a prediction interval for the clutch size for a salamander with snout-vent length of 205 based on available data is not recommended.

13.43 **a.**

$$b = \dfrac{1081.5 - \left[\dfrac{(269)(51)}{14}\right]}{7445 - \left[\dfrac{(269)^2}{14}\right]} = \dfrac{101.571}{2276.357} = 0.04462$$

a = 3.6429   (0.04462)(19.214) = 2.78551
The equation of the estimated regression line is $\hat{y}$ = 2.78551 + 0.04462x.

b. H_o: = 0 H_a: $\neq$ 0

= 0.05

The test statistic is: $t = \dfrac{b}{s_b}$ with df. = 12.

SSResid = 190.78   (2.78551)(51)   (0.00462)(1081.5) = 0.46246

$$s_e^2 = \dfrac{0.46246}{12} = 0.0385$$

$$s_b^2 = \dfrac{0.0385}{2276.357} = 0.0000169, \quad s_b = 0.004113$$

$$t = \frac{0.04462}{0.004113} = 10.85$$

P-value = 2(area under the 12 df t curve to the right of 10.85) = 2(0) = 0.
Since the P-value is less than , the null hypothesis is rejected. The data suggests that the simple linear regression model provides useful information for predicting moisture content from knowledge of time.

c. The point estimate of the moisture content of an individual box that has been on the shelf 30 days is 2.78551 + 0.04462(30) = 4.124.
The 95% prediction interval is

$$4.124 \pm (2.18)\sqrt{0.0385}\sqrt{1 + \frac{1}{14} + \frac{(30 - 19.214)^2}{2276.357}}$$

4.124   2.18(0.2079)   4.124   0.453 = (3.671, 4.577).

d. Since values greater than equal to 4.1 are included in the interval constructed in **c**, it is very plausible that a box of cereal that has been on the shelf 30 days will not be acceptable.

13.45 a.

$$b = \frac{57760 - \left[\dfrac{(1350)(600)}{15}\right]}{155400 - \left[\dfrac{(1350)^2}{15}\right]} = \frac{3760}{33900} = 0.1109$$

a = 40   (0.1109)(90) = 30.019
The equation for the estimated regression line is $\hat{y} = 30.019 + 0.1109x$.

b. When x = 100, the point estimate of   +   (100) is   30.019 + 0.1109(100) = 41.109.
SSResid = 24869.33   (30.019)(600)   (0.1109)(57760) = 452.346

$$s_e^2 = \frac{452.346}{13} = 34.7958$$

$$s_{a+b(100)}^2 = 34.7958\left[\frac{1}{15} + \frac{(100 - 90)^2}{33900}\right] = 2.422$$

$$s_{a+b(100)} = \sqrt{2.422} = 1.5564$$

The 90% confidence interval for the mean blood level for people who work where the air lead level is 100 is 41.109   (1.77)(1.5564)   41.109   2.755   (38.354, 43.864).

c. The prediction interval is 41.109   (1.77)$\sqrt{34.7958 + 2.422}$ 41.109   10.798   (30.311, 51.907).

d. The interval of part **b** is for the mean blood level of all people who work where the air lead level is 100. The interval of part **c** is for a single randomly selected individual who works where the air lead level is 100.

13.47 **a.** The 95% prediction interval for sunburn index when distance is 35 is 2.5225

$2.16 \sqrt{(0.25465)^2 + (0.07026)^2}$ 2.5225   0.5706   (1.9519, 3.0931).

The 95% prediction interval for sunburn index when distance is 45 is 1.9575

$2.16 \sqrt{(0.25465)^2 + (0.06857)^2}$ 1.9575   0.5696   (1.3879, 2.5271).

The pair of intervals form a set of simultaneous prediction intervals with prediction level of at least [100   2(5)]% = 90%.

b. The simultaneous prediction level would be at least [100   3(1)]% = 97%.

Exercises 13.49 – 13.57

13.49 The quantity r is a statistic as its value is calculated from the sample. It is a measure of how strongly the sample x and y values are linearly related. The value of r is an estimate of .
The quantity   is a population characteristic. It measures the strength of linear relationship between the x and y values in the population.

13.51 Let   denote the true correlation coefficient between teaching evaluation index and annual raise.

H_o: = 0 H_a: 0

= 0.05

$$t = \dfrac{r}{\sqrt{\dfrac{1-r^2}{n-2}}}$$ with df. = 351   n = 353, r = 0.11

$$t = \dfrac{0.11}{\sqrt{\dfrac{1-(0.11)^2}{351}}} = \dfrac{0.11}{0.05305} = 2.07$$

The t curve with 351 df is essentially the z curve.

P-value = 2(area under the z curve to the right of 2.07) = 2(0.0192) = 0.0384.

Since the P-value is less than   , H_o is rejected. There is sufficient evidence in the sample to conclude that there appears to be a linear association between teaching evaluation index and annual raise.

According to the guidelines given in the text book, r = 0.11 suggests only a weak linear relationship.

Since r^2 = 0.0121, fitting the simple linear regression model to the data would result in only about 1.21% of observed variation in annual raise being explained.

13.53 **a.** Let ρ denote the correlation coefficient between time spent watching television and grade point average in the population from which the observations were selected.

$$H_o: \rho = 0 \qquad H_a: \rho < 0$$
$$\alpha = 0.01$$

$$t = \frac{r}{\sqrt{\dfrac{1-r^2}{n-2}}} \qquad \text{with df.} = 526$$

$n = 528, \; r = -0.26$

$$t = \frac{-0.26}{\sqrt{\dfrac{1-(-0.26)^2}{526}}} = \frac{-0.26}{0.042103} = -6.175$$

The t curve with 526 df is essentially the z curve.

P-value = area under the z curve to the left $-6.175 \approx 0$.

Since the P-value is less than α, H_o is rejected. The data does support the conclusion that there is a negative correlation in the population between the two variables, time spent watching television and grade point average.

b. The coefficient of determination measures the proportion of observed variation in grade point average explained by the regression on time spent watching television. This value would be $(-0.26)^2 = 0.0676$. Thus only 6.76% of the observed variation in grade point average would be explained by the regression. This is not a substantial percentage.

13.55 From the summary quantities:

$$S_{xy} = 673.65 - \left[\frac{(136.02)(39.35)}{9}\right] = 78.94$$

$$S_{xx} = 3602.65 - \left[\frac{(136.02)^2}{9}\right] = 1546.93$$

$$S_{yy} = 184.27 - \left[\frac{(39.35)^2}{9}\right] = 12.223$$

$$r = \frac{78.94}{\sqrt{(1546.93)(12.223)}} = \frac{78.94}{137.51} = 0.574$$

Let ρ denote the correlation between surface and subsurface concentration.

$$H_o: \rho = 0 \qquad H_a: \rho \neq 0$$
$$\alpha = 0.05$$

$$t = \frac{r}{\sqrt{\dfrac{1-r^2}{n-2}}} \qquad \text{with df.} = 7$$

$$t = \frac{0.574}{\sqrt{\dfrac{1-(0.574)^2}{7}}} = 1.855$$

P-value = 2(area under the 7 df t curve to the right of 1.855)   2(0.053) = 0.106.
Since the P-value exceeds   , H_o is not rejected. The data does not support the conclusion that there is a linear relationship between surface and subsurface concentration.

13.57 H_o: = 0 H_a: 0
        = 0.05

$$t = \frac{r}{\sqrt{\dfrac{1-r^2}{n-2}}} \quad \text{with df.} = 9998$$

n = 10000, r = 0.022

$$t = \frac{0.022}{\sqrt{\dfrac{1-(0.022)^2}{9998}}} = 2.2$$

The t curve with 9998 df is essentially the z curve.
P-value = 2(area under the z curve to the right of 2.2) = 2(0.0139) = 0.0278.
Since the P-value is less than   , H_o is rejected. The results are statistically significant. Because of the extremely large sample size, it is easy to detect a value of   which differs from zero by a small amount. If   is very close to zero, but not zero, the practical significance of a non-zero correlation may be of little consequence.

Exercises 13.59 – 13.75

13.59 **a.** $t = \dfrac{-0.18}{\sqrt{\dfrac{1-(-.18)^2}{345}}} = \dfrac{-0.18}{0.052959} = -3.40$ with df = 345

If the test was a one-sided test, then the P-value equals the area under the z curve to the left of 3.40, which is equal to 0.0003. It the test was a two-sided test, then the P-value is 2(0.0003) = 0.0006. While the researchers' statement is true, they could have been more precise in their statement about the P-value.
b. From my limited experience, I have observed that the more visible a person's sense of humor, the less depressed they *appear* to be. This would suggest a negative correlation between Coping Humor Scale and Sense of Humor.
c. Since $r^2 = (\ 0.18)^2 = 0.0324$, only about 3.24% of the observed variability in sense of humor can be explained by the linear regression model. This suggests that a simple linear regression model may not give accurate predictions.

13.61   **a.** H_o: $\beta = 0$ H_a: $\beta \neq 0$

$\alpha = 0.01$

$$t = \frac{b}{s_b}$$

From the Minitab output, t = 3.95 and the P-value = 0.003. Since the P-value is less than α, H_o is rejected. The data supports the conclusion that the simple linear regression model is useful.

b. A 95% confidence interval for β is 2.3335 ± 2.26(0.5911) → 2.3335 ± 1.3359 → (3.6694, 0.9976).

c. a+b(10) = 88.796 − 2.3335(10) = 65.461

$s_{a+b(10)} = 0.689$

The 95% prediction interval for an individual y when x = 10 is

$65.461 \pm 2.26\sqrt{(0.689)^2 + 4.789}$ → 65.461 ± 5.185 → (60.276, 70.646).

d. Because x = 11 is farther from $\overline{x}$ than x = 10 is from $\overline{x}$.

13.63   **a.** Let ρ denote the correlation coefficient between soil hardness and trail length.

H_o: $\rho = 0$ H_a: $\rho < 0$

$\alpha = 0.05$

$$t = \frac{r}{\sqrt{\dfrac{1-r^2}{(n-2)}}} \quad \text{with df.} = 59$$

$$t = \frac{-0.6213}{\sqrt{\dfrac{1-(-0.6213)^2}{59}}} = -6.09$$

P-value = area under the 59 df t curve to the left of −6.09 ≈ 0.

Since the P-value is less than α, the null hypothesis is rejected. The data supports the conclusion of a negative correlation between trail length and soil hardness.

b. When x* = 6, a+b(6) = 11.607 − 1.4187(6) = 3.0948

$$s^2_{a+b(6)} = (2.35)^2 \left[\frac{1}{61} + \frac{(6-4.5)^2}{250} \right] = 0.1402$$

$s_{a+b(6)} = \sqrt{0.1402} = 0.3744$

The 95% confidence interval for the mean trail length when soil hardness is 6 is

3.0948 ± 2.00(0.3744) → 3.0948 ± 0.7488 → (2.346, 3.844).

c. When x* = 10, a+b(10) = 11.607 − 1.4187(10) = −2.58

According to the least-squares line, the predicted trail length when soil hardness is 10 is −2.58. Since trail length cannot be negative, the predicted value makes no sense. Therefore, one would not use the simple linear regression model to predict trail length when hardness is 10.

13.65 $n = 17$, $x = 821$, $x^2 = 43447$, $y = 873$, $y^2 = 46273$, $xy = 40465$,

$$S_{xy} = 40465 - \left[\frac{(821)(873)}{17}\right] = 40465 - 42160.7647 = -1695.7647$$

$$S_{xx} = 43447 - \left[\frac{(821)^2}{17}\right] = 43447 - 39649.4706 = 3797.5294$$

$$S_{yy} = 46273 - \left[\frac{(873)^2}{17}\right] = 46273 - 44831.1176 = 1441.8824$$

$$b = \frac{-1695.7647}{3797.5294} = -0.4465$$

$a = 51.3529$ (0.4465)(48.2941) = 72.9162

SSResid = 46273 72.9162(873) (0.4465)(40465) = 684.78

$$s_e^2 = \frac{684.78}{15} = 45.652 , \quad s_e = 6.7566$$

$$s_b = \frac{6.7566}{\sqrt{3797.5294}} = 0.1096$$

a. Let   denote the average change in percentage area associated with a one year increase in age.

$\qquad H_o$: $= 0.5$ H_a: $\neq 0.5$

$\qquad\quad = 0.10$

$$t = \frac{b-(-0.5)}{s_b} \quad \text{with df.} = 15$$

$$t = \frac{-0.4465-(-0.5)}{0.1096} = 0.49$$

P-value = 2(area under the 15 df t curve to the right of 0.49) 2(0.312) = 0.624.
Since the P-value exceeds   , H_0 is not rejected. There is not sufficient evidence in the sample to contradict the prior belief of the researchers.

b. When $x^* = 50$, $\hat{y} = 72.9162 + ($ 0.4465)(50) = 50.591

$$s_{a+b(50)} = 6.7471\sqrt{\frac{1}{17} + \frac{(50-48.2941)^2}{3797.5294}} = 1.647$$

The 95% confidence interval for the true average percent area covered by pores for all 50 year-olds is

50.591 (2.13)(1.647)   50.591 3.508   (47.083, 54.099).

13.67 Summary values for Leptodactylus ocellatus: $n = 9$, $x = 64.2$, $x^2 = 500.78$, $y = 19.6$, $y^2 = 47.28$, $xy = 153.36$

From these: $b = 0.31636$, SSResid $= 0.3099$, $(x \quad \bar{x})^2 \quad 42.82$

Summary values for Bufa marinus: $n = 8$, $x = 55.9$, $x^2 = 425.15$, $y = 21.6$, $y^2 = 62.92$, $xy = 163.63$

From these: $b = 0.35978$, SSResid $= 0.1279$, $(x \quad \bar{x})^2 \quad 34.549$

$$s^2 = \frac{0.3099 + 0.1279}{9 + 8 - 4} = \frac{0.4378}{13} = 0.0337$$

H_o: H_a:

$= 0.05$

The test statistic is: $t = \dfrac{b - b'}{\sqrt{\dfrac{s^2}{SS_x} + \dfrac{s^2}{SS'_x}}}$ with df. $= 9 + 8 \quad 4 = 13$.

$$t = \frac{0.31636 - 0.35978}{\sqrt{\dfrac{0.0337}{42.82} + \dfrac{0.0337}{34.549}}} = \frac{-0.04342}{0.04198} = -1.03$$

P-value $= 2$(area under the 13 df t curve to the left of 1.03) $= 2(0.161) = 0.322$.

Since the P-value exceeds , the null hypothesis of equal regression slopes cannot be rejected. The data are consistent with the hypothesis that the slopes of the true regression lines for the two different frog populations are identical.

13.69 When the point is included in the computations, the slope will be negative and much more extreme (farther from 0) than if the point is excluded from the computations. Changing the slope will also have an effect on the intercept.

13.71 Since the P-value of 0.0076 is smaller than most reasonable levels of significance, the conclusion of the model utility test would be that the percentage raise does appear to be linearly related to productivity. This should be confirmed by examining a scatter plot of percentage raise versus productivity.

13.73 Even though the P-value is small and $r^2 = 0.65$, the variability about the least-squares line appears to increase as soil depth increases. If this is so, a residual plot would be "funnel shape", opening up toward the right. One of the conditions for valid application of the procedures described in this chapter is that the variability of the y's be constant. It appears that this requirement may not be true in this instance.

13.75 a. When $r = 0$, then $s_e \quad s_y$. The least squares line in this case is a horizontal line with intercept of $\bar{y}$.

b. When r is close to 1 in absolute value, then s_e will be much smaller than s_y.

Chapter 14

Exercises 14.1 – 14.13

14.1 A deterministic model does not have the random deviation component e, while a probabilistic model does contain such a component.

Let   y =   total number of goods purchased at a service station which sells only one grade of gas and one type of motor oil.

   x_1 =   gallons of gasoline purchased

   x_2 =   number of quarts of motor oil purchased.

Then y is related to x_1 and x_2 in a deterministic fashion.

Let   y =   IQ of a child

   x_1 =   age of the child

   x_2 =   total years of education of the parents.

Then y is related to x_1 and x_2 in a probabilistic fashion.

14.3 The following multiple regression model is suggested by the given statement.

$$y = \beta_0 + \beta_1 x_1 + \beta_2 x_2 + e.$$

An interaction term is not included in the model because it is given that x_1 and x_2 make independent contributions to academic achievement.

14.5 **a.** 415.11 – 6.6(20) – 4.5(40) = 103.11

b. 415.11 – 6.6(18.9) – 4.5(43) = 96.87

c. β_1 = – 6.6. Hence 6.6 is the expected *decrease* in yield associated with a one-unit increase in mean temperature between date of coming into hop and date of picking when the mean percentage of sunshine remains fixed.

β_2 = – 4.5. So 4.5 is the expected *decrease* in yield associated with a one-unit increase in mean percentage of sunshine when mean temperature remains fixed.

14.7 **a.**

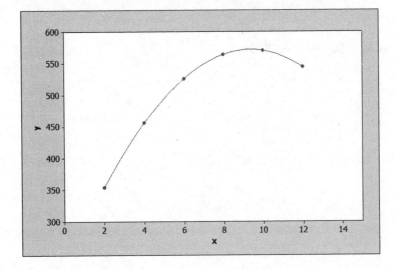

b. The mean chlorine content at x = 8 is 564, while at x = 10 it is 570. So the mean chlorine content is higher for x = 10 than for x = 8.

c. When x = 9, $y = 220 + 75(9) - 4(9)^2 = 571$.

The change in mean chlorine content when the degree of delignification increases from 8 to 9 is 571 – 564 = 7.

The change in mean chlorine content when the degree of delignification increases from 9 to 10 is 570 – 571 = –1.

14.9　　**a.**

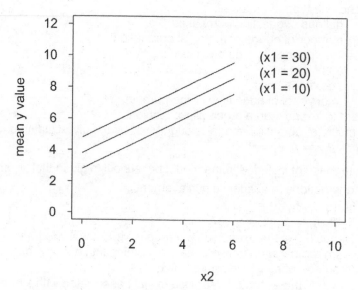

For $x_1 = 30$, $y = 4.8 + 0.8x_2$

For $x_1 = 20$, $y = 3.8 + 0.8x_2$

For $x_1 = 10$, $y = 2.8 + 0.8x_2$

b.

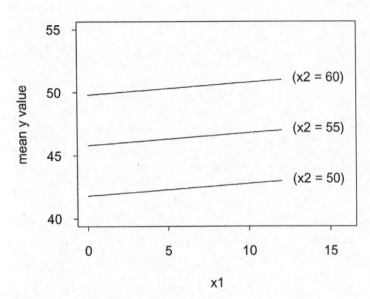

For $x_2 = 60$, $y = 49.8 + 0.1x_2$
For $x_2 = 55$, $y = 45.8 + 0.1x_2$
For $x_2 = 50$, $y = 41.8 + 0.1x_2$

c. The parallel lines in each graph are attributable to the lack of interaction between the two independent variables.

d.

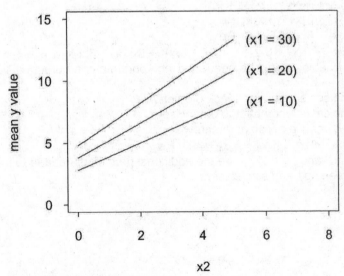

For $x_1 = 30$, $y = 4.8 + 1.7x_2$
For $x_1 = 20$, $y = 3.8 + 1.4x_2$
For $x_1 = 10$, $y = 2.8 + 1.1x_2$

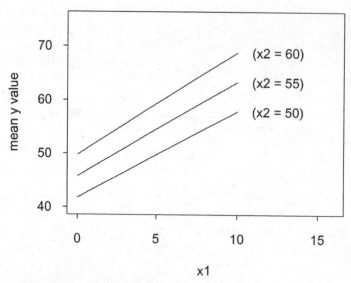

For $x_2 = 60$, $y = 49.8 + 1.9x_2$
For $x_2 = 55$, $y = 45.8 + 1.75x_2$
For $x_2 = 50$, $y = 41.8 + 1.6x_2$

Because there is an interaction term, the lines are not parallel.

14.11 **a.** $y = \alpha + \beta_1x_1 + \beta_2x_2 + \beta_3x_3 + e$
b. $y = \alpha + \beta_1x_1 + \beta_2x_2 + \beta_3x_3 + \beta_4x_1^2 + \beta_5x_2^2 + \beta_6x_3^2 + e$
c. $y = \alpha + \beta_1x_1 + \beta_2x_2 + \beta_3x_3 + \beta_4x_1x_2 + e$
$y = \alpha + \beta_1x_1 + \beta_2x_2 + \beta_3x_3 + \beta_4x_1x_3 + e$
$y = \alpha + \beta_1x_1 + \beta_2x_2 + \beta_3x_3 + \beta_4x_2x_3 + e$
d. $y = \alpha + \beta_1x_1 + \beta_2x_2 + \beta_3x_3 + \beta_4x_1^2 + \beta_5x_2^2 + \beta_6x_3^2 + \beta_7x_1x_2 + \beta_8x_1x_3 + \beta_9x_2x_3 + e$

14.13 **a.** Three dummy variables would be needed to incorporate a non-numerical variable with four categories.
$x_3 = 1$ if the car is a sub-compact, 0 otherwise
$x_4 = 1$ if the car is a compact, 0 otherwise
$x_5 = 1$ if the car is a midsize, 0 otherwise
$y = \alpha + \beta_1x_1 + \beta_2x_2 + \beta_3x_3 + \beta_4x_4 + \beta_5x_5 + e$
b. $x_6 = x_1x_3$, $x_7 = x_1x_4$, and $x_8 = x_1x_5$ are the additional predictors needed to incorporate interaction between age and size class.

Exercises 14.15 – 14.33

14.15 **a.** $b_1 = -2.18$ is the estimated change in mean fish intake associated with a one-unit increase in water temperature when the value of the other predictor variables are fixed. Since the sign of b_1 negative, the change is actually a decrease.

$b_4 = 2.32$ is the estimated change (increase) in mean fish intake associated with a one-unit increase in speed   when the value of the other predictor variables are fixed.

b. $R^2 = 1 - \dfrac{\text{SSResid}}{\text{SSTo}} = \dfrac{\text{SSRegr}}{\text{SSTo}} = \dfrac{1486.9}{1486.9 + 2230.2} = \dfrac{1486.9}{3717.1} = .4$

c. $s_e^2 = \dfrac{\text{SSResid}}{n - (k+1)} = \dfrac{2230.2}{26 - 5} = \dfrac{2230.2}{21} = 106.2$

$s_e = \sqrt{106.20} = 10.305$

d. Adjusted $R^2 = 1 - \left[\dfrac{n-1}{n-(k+1)}\right]\dfrac{\text{SSResid}}{\text{SSTo}} = 1 - \left(\dfrac{25}{21}\right)\left(\dfrac{2230.2}{3717.1}\right) = 1 - .7143 = .2857$

Adjusted R^2 has a smaller value than R^2.

14.17 **a.** $df_1 = k = 2$ $df_2 = n - (k+1) = 21 - 3 = 18$;   P-value > 0.10
b. $df_1 = k = 8$ $df_2 = n - (k+1) = 25 - 9 = 16$;   0.01 > P-value > 0.001
c. $df_1 = k = 5$ $df_2 = n - (k+1) = 26 - 6 = 20$;   0.05 > P-value > 0.01
d. $df_1 = k = 5$ $df_2 = n - (k+1) = 20 - 6 = 14$;   0.001 > P-value
e. $df_1 = k = 5$ $df_2 = n - (k+1) = 100 - 6 = 94$;   Using $df_2 = 90$, 0.05 > P-value > 0.01

14.19 The fitted model was $y = \alpha + \beta_1x_1 + \beta_2x_2 + \beta_3x_3 + \beta_4x_4 + \beta_5x_5 + \beta_6x_6 + \beta_7x_7 + e$
$H_0: \beta_1 = \beta_2 = \beta_3 = \beta_4 = \beta_5 = \beta_6 = \beta_7 = 0$
H_a: at least one among β_1, β_2, β_3, β_4, β_5, β_6, β_7 is not zero
$\alpha = 0.05$

$F = \dfrac{R^2/k}{(1-R^2)/[n-(k+1)]}$

$n = 210$, $df_1 = k = 7$, $df_2 = n - (k+1) = 210 - 8 = 202$
$R^2 = 0.543$

$F = \dfrac{R^2/k}{(1-R^2)/[n-(k+1)]} = \dfrac{.543/7}{(1-.543)/202} = 34.286$

From Appendix Table 6, 0.001 > P-value.
Since the P-value is less than α, the null hypothesis is rejected. There does appear to be a useful linear relationship between y and at least one of the seven predictors.

14.21 **a.** Estimated mean value of $y = 86.85 - 0.12297x_1 + 5.090x_2 - 0.07092x_3 + 0.001538x_4$
b. $H_0: \beta_1 = \beta_2 = \beta_3 = \beta_4 = 0$
H_a: at least one among β_1, β_2, β_3, β_4, is not zero
$\alpha = 0.01$

$F = \dfrac{R^2/k}{(1-R^2)/[n-(k+1)]}$

$$n = 31, \quad df_1 = k = 4, \quad df_2 = n - (k + 1) = 31 - 5 = 26$$
$$R^2 = 0.908$$

$$F = \frac{R^2/k}{(1-R^2)/[n-(k+1)]} = \frac{.908/4}{(1-.908)/26} = 64.15$$

From Appendix Table 6, 0.001 > P-value

Since the P-value is less than α, the null hypothesis is rejected. There does appear to be a useful linear relationship between y and at least one of the four predictors.

c. $R^2 = 0.908$. This means that 90.8% of the variation in the observed tar content values has been explained by the fitted model.

$s_e = 4.784$. This means that the typical distance of an observation from the corresponding mean value is 4.784.

14.23 **a.** The regression model being fitted is $y = \alpha + \beta_1 x_1 + \beta_2 x_2 + e$.

Using MINITAB, the regression command yields the following output.

The regression equation is

weight(g) = - 511 + 3.06 length(mm) - 1.11 age(years)

Predictor	Coef	SE Coef	T	P
Constant	-510.9	286.1	-1.79	0.096
length(mm)	3.0633	0.8254	3.71	0.002
age(years)	-1.113	9.040	-0.12	0.904

S = 94.24   R-Sq = 59.3%   R-Sq(adj) = 53.5%

Analysis of Variance

Source	DF	SS	MS	F	P
Regression	2	181364	90682	10.21	0.002
Residual Error	14	124331	8881		
Total	16	305695			

b. $H_o: \beta_1 = \beta_2 = 0$

H_a: At least one of the two β_i's is not zero.

$$F = \frac{SSRegr/k}{SSResid/[n-(k+1)]}$$

$$n = 17, \quad df_1 = k = 2, \quad df_2 = n - (k + 1) = 17 - 3 = 14$$

$$F = \frac{SSRegr/k}{SSResid/[n-(k+1)]} = \frac{181364/2}{124331/14} = 10.21$$

From the Minitab output, the P-value = 0.002. Since the P-value is less than 0.05 (we have chosen $\alpha = 0.05$ for illustration) the null hypothesis is rejected. The data suggests that the multiple regression model is useful for predicting weight.

14.25 **a.** Using MINITAB to fit the required regression model yields the following output.

The regression equation is
volume = - 859 + 23.7 minwidth + 226 maxwidth + 225 elongation

Predictor	Coef	SE Coef	T	P
Constant	-859.2	272.9	-3.15	0.005
Minwidth	23.72	85.66	0.28	0.784
Maxwidth	225.81	85.76	2.63	0.015
Elongation	225.24	90.65	2.48	0.021

$S = 287.0$ R-Sq = 67.6%   R-Sq(adj) = 63.4%

Analysis of Variance

Source	DF	SS	MS	F	P
Regression	3	3960700	1320233	16.03	0.000
Residual Error	23	1894141	82354		
Total	26	5854841			

b. Adjusted R^2 takes into account the number of predictors used in the model whereas R^2 does not do so. In particular, adjusted R^2 enables us to make a "fair comparison" of the performances of models with differing numbers of predictors.

c. We test

$H_o: \beta_1 = \beta_2 = \beta_3 = 0$
H_a: At least one of the three β_i's is not zero.
$\alpha = 0.05$ (for illustration)

$$F = \frac{R^2/k}{(1 - R^2)/[n - (k + 1)]}$$

$n = 27$, $df_1 = k = 3$, $df_2 = n - (k + 1) = 27 - 4 = 23$
$R^2 = 0.676$

$$F = \frac{R^2/k}{(1 - R^2)/[n - (k + 1)]} = \frac{0.676/3}{(1 - 0.676)/23} = 16.03$$

The corresponding P-value is 0.000 (correct to 3 decimals). Since the P-value is less than α, the null hypothesis is rejected. There does appear to be a useful linear relationship between y and at least one of the three predictors.

14.27 **a.** SSResid = 390.4347
SSTo = 7855.37 − 14(21.1071)² = 1618.2093
SSRegr = 1618.2093 − 390.4347 = 1227.7746

b. $R^2 = \dfrac{1227.7746}{1618.2093} = .759$

This means that 75.9 percent of the variation in the observed shear strength values has been explained by the fitted model.

c. H_o: $\beta_1 = \beta_2 = \beta_3 = \beta_4 = \beta_5 = 0$

H_a: at least one among β_1, β_2, β_3, β_4, β_5, is not zero

$\alpha = 0.05$

$$F = \frac{R^2/k}{(1-R^2)/[n-(k+1)]}$$

$n = 14$, $df_1 = k = 5$, $df_2 = n - (k+1) = 14 - 6 = 8$

$R^2 = 0.759$

$$F = \frac{R^2/k}{(1-R^2)/[n-(k+1)]} = \frac{.759/5}{(1-.759)/8} = 5.039$$

From Appendix Table 6, 0.05 > P-value >0.01.

Since the P-value is less than α, the null hypothesis is rejected. There does appear to be a useful linear relationship between y and at least one of the predictors. The data suggests that the independent variables as a group do provide information that is useful for predicting shear strength.

14.29 H_o: $\beta_1 = \beta_2 = 0$

H_a: At least one of the two β_i's is not zero.

$\alpha = 0.01$

$$F = \frac{R^2/k}{(1-R^2)/[n-(k+1)]}$$

$n = 24$, $df_1 = k = 2$, $df_2 = n - (k+1) = 24 - 3 = 21$

$$F = \frac{R^2/k}{(1-R^2)/[n-(k+1)]} = \frac{.902/2}{(1-.902)/21} = 96.64$$

From Appendix Table 6, 0.001 > P-value.

Since the P-value is less than α, the null hypothesis is rejected. The data suggests that the quadratic model does have utility for predicting yield.

14.31 Using MINITAB, the regression command yields the following output.

The regression equation is

Y = − 151 − 16.2 X1 + 13.5 X2 + 0.0935 X1-SQ − 0.253 X2-SQ + 0.0492 X1*X2.

Predictor	Coef	Stdev	t-ratio	p
Constant	−151.4	134.1	−1.13	0.292
X1	−16.216	8.831	−1.84	0.104
X2	13.476	8.187	1.65	0.138
X1-SQ	0.09353	0.07093	1.32	0.224
X2-SQ	−0.2528	0.1271	−1.99	0.082
X1*X2	0.4922	0.2281	2.16	0.063

It can be seen (except for differences due to rounding errors) that the estimated regression equation given in the problem is correct.

14.33 Using MINITAB, the regression command yields the following output.

The regression equation is INF_RATE = 35.8 − 0.68 AVE_TEMP + 1.28 AVE_RH

Predictor	Coef	Stdev	t-ratio	p
Constant	35.83	53.54	0.67	0.508
AVE_TEMP	−0.676	1.436	−0.47	0.641
AVE_RH	1.2811	0.4243	3.02	0.005

s = 22.98   R-sq = 55.0%   R-sq(adj) = 52.1%

Analysis of Variance

SOURCE	DF	SS	MS	F	p
Regression	2	20008	10004	18.95	0.000
Error	31	16369	528		
Total	33	36377			

H_o: $\beta_1 = \beta_2 = 0$
H_a: At least one of the two β_i's is not zero.

$$F = \frac{SSRegr/k}{SSResid/[n-(k+1)]}$$

$n = 34$, $df_1 = k = 2$, $df_2 = n - (k + 1) = 34 - 3 = 31$

$$F = \frac{SSRegr/k}{SSResid/[n-(k+1)]} = \frac{20008/2}{16369/31} = 18.95$$

From the Minitab output, the P-value ≈ 0. Since the P-value is less than α, the null hypothesis is rejected. The data suggests that the multiple regression model has utility for predicting infestation rate.

Exercises 14.35 – 14.47

14.35   **a.** The degrees of freedom for error is 100 − (7 + 1) = 92. From Appendix Table 3, the critical t value is approximately 1.99.
The 95% confidence interval for β_3 is −0.489 ± (1.99)(0.1044) ⟹ −4.89 ± 0.208 ⟹ (−0.697, −0.281).
With 95% confidence, the change in the mean value of a vacant lot associated with a one unit increase in distance from the city's major east-west thoroughfare is a decrease of as little as 0.281 or as much as 0.697.

b. H_o: $\beta_1 = 0$ H_a: $\beta_1 \neq 0$
$\alpha = 0.05$

$t = \dfrac{b_1}{s_{b_1}}$ with df. = 92

$t = \dfrac{-.183}{.3055} = -0.599$

P-value = 2(area under the 92 df t curve to the left of –0.599) ≈ 2(0.275) = 0.550.
Since the P-value exceeds α, the null hypotheses is not rejected. This means that there is not sufficient evidence to conclude that there is a difference in the mean value of vacant lots that are zoned for residential use and those that are not zoned for residential use.

14.37 **a.** H_o: $\beta_1 = \beta_2 = 0$

H_a: At least one of the two β_i's is not zero.

$\alpha = 0.05$

$$F = \frac{R^2/k}{(1-R^2)/[n-(k+1)]}$$

$n = 50$, $df_1 = k = 2$, $df_2 = n - (k + 1) = 50 - 3 = 47$, $R^2 = 0.86$

$$F = \frac{R^2/k}{(1-R^2)/[n-(k+1)]} = \frac{.86/2}{(1-.86)/47} = 144.36$$

From Appendix Table 6, 0.001 > P-value.

Since the P-value is less than α, the null hypothesis is rejected. The data suggests that the quadratic regression model has utility for predicting MDH activity.

b. H_o: $\beta_2 = 0$ H_a: $\beta_2 \neq 0$

$\alpha = 0.01$

$t = \dfrac{b_2}{s_{b_2}}$ with df. = 47

$t = \dfrac{.0446}{.0103} = 4.33$

P-value = 2(area under the 47 df t curve to the right of 4.33) ≈ 2(0) = 0.

Since the P-value is less than α, the null hypothesis is rejected. The quadratic term is an important term in this model.

c. The point estimate of the mean value of MDH activity for an electrical conductivity level of 40 is $-0.1838 + 0.0272(40) + 0.0446(40^2) = -0.1838 + 0.0272(40) + 0.0446(1600) = 72.2642$.

The 90% confidence interval for the mean value of MDH activity for an electrical conductivity level of 40 is $72.2642 \pm (1.68)(0.120) \Rightarrow 72.2642 \pm 0.2016 \Rightarrow (72.0626, 72.4658)$

14.39 **a.** The value 0.469 is an estimate of the expected change (increase) in the mean score of students associated with a one unit increase in the student's expected score holding time spent studying and student's grade point average constant.

b. H_o: $\beta_1 = \beta_2 = \beta_3 = 0$

H_a: At least one of the three β_i's is not zero.

$\alpha = 0.05$

$$F = \frac{R^2/k}{(1-R^2)/[n-(k+1)]}$$

$n = 107$, $df_1 = k = 3$, $df_2 = n - (k + 1) = 107 - 4 = 103$, $R^2 = 0.686$

$$F = \frac{R^2/k}{(1-R^2)/[n-(k+1)]} = \frac{.686/3}{(1-.686)/103} = 75.01$$

From Appendix Table 6, 0.001 > P-value.

Since the P-value is less than α, the null hypothesis is rejected. The data suggests that there is a useful linear relationship between exam score and at least one of the three predictor variables.

c. The 95% confidence interval for β_2 is $3.369 \pm (1.98)(0.456) \Rightarrow 3.369 \pm 0.903 \Rightarrow (2.466, 4.272)$.

d. The point prediction would be $2.178 + 0.469(75) + 3.369(8) + 3.054(2.8) = 72.856$.

e. The prediction interval would be $72.856 \pm (1.98) \sqrt{s_e^2 + (1.2)^2}$.

To determine s_e^2, proceed as follows. From the definition of R^2, it follows that SSResid = $(1-R^2)$SSTo. So SSResid = $(1-0.686)(10200) = 3202.8$.

Then, $s_e^2 = \dfrac{3202.8}{103} = 31.095$.

The prediction interval becomes

$72.856 \pm (1.98) \sqrt{31.095 + (1.2)^2} \Rightarrow 72.856 \pm (1.98)(5.704)$

$\Rightarrow 72.856 \pm 11.294 \Rightarrow (61.562, 84.150)$.

14.41 H_o: $\beta_3 = 0$ H_a: $\beta_3 \neq 0$

$\alpha = 0.05$

$t = \dfrac{b_3}{s_{b_3}}$ with d.f. = 363

$t = \dfrac{.00002}{.000009} = 2.22$

P-value = 2(area under the 363 df t curve to the right of 2.22) $\approx$ 2(0.014) = 0.028.

Since the P-value is less than α, the null hypothesis is rejected. The conclusion is that the inclusion of the interaction term is important.

14.43 **a.** H_o: $\beta_1 = \beta_2 = \beta_3 = 0$

H_a: At least one of the three β_i's is not zero.

$\alpha = 0.05$

Test statistic: $F = \dfrac{\text{SSRegr}/k}{\text{SSResid}/[n-(k+1)]}$

$F = \dfrac{5073.4/3}{1854.1/6} = 5.47$

From Appendix Table 6, 0.05 > P-value > 0.01.

Since the P-value is less than α, the null hypothesis is rejected. The data suggests that the model has utility for predicting discharge amount.

b. H_o: $\beta_3 = 0$ H_a: $\beta_3 \neq 0$

$\alpha = 0.05$

The test statistic is: $t = \dfrac{b_3}{s_{b_3}}$ with df. = 6.

$$t = \frac{8.4}{199} = 0.04$$

P-value = 2(area under the 6 df t curve to the right of 0.04) ≈ 2(0.48) = 0.96.
Since the P-value exceeds α, the null hypothesis is not rejected. The data suggests that the interaction term is not needed in the model, if the other two independent variables are in the model.

c. No. The model utility test is testing all variables simultaneously (that is, as a group). The t test is testing the contribution of an individual predictor when used in the presence of the remaining predictors. Results indicate that, given two out of the three predictors are included in the model, the third predictor may not be necessary.

14.45 The point prediction for mean phosphate adsorption when $x_1 = 160$ and $x_2 = 39$ is at the midpoint of the given interval. So the value of the point prediction is (21.40 + 27.20)/2 = 24.3. The t critical value for a 95% confidence interval is 2.23. The standard error for the point prediction is equal to (27.20 − 21.40)/2(2.23) = 1.30. The t critical value for a 99% confidence interval is 3.17. Therefore, the 99% confidence interval would be 24.3 ± (3.17)(1.3) ⇒ 24.3 ± 4.121 ⇒ (20.179, 28.421).

a.

The regression equation is
Weight = - 511 + 3.06 Length - 1.11 Age

Predictor	Coef	SE Coef	T	P
Constant	-510.9	286.1	-1.79	0.096
Length	3.0633	0.8254	3.71	0.002
Age	-1.113	9.040	-0.12	0.904

t = 3.71, P-value = .002, reject H_0, length cannot be eliminated.
t = -0.12, P-value = 0.904, fail to reject H_0, age could be eliminated.

b.

The regression equation is
Weight = - 1516 + 5.71 Length + 2.48 Age - 238 Year

Predictor	Coef	SE Coef	T	P
Constant	-1515.8	290.8	-5.21	0.000
Length	5.7149	0.7972	7.17	0.000
Age	2.479	5.925	0.42	0.683
Year	-237.95	52.96	-4.49	0.001

t = -4.49, P-value = .001, reject H_0, year is a useful predictor.

14.47 **a.** H_o: $\beta_1 = \beta_2 = 0$

H_a: At least one of the two β_i's is not zero.

$\alpha = 0.05$

Test statistic: $F = \dfrac{\text{SSRegr} / k}{\text{SSResid} / [n-(k+1)]}$

$F = \dfrac{237.52/2}{26.98/7} = 30.81$

From Appendix Table 6, 0.001 > P-value.

Since the P-value is less than α, the null hypothesis is rejected. The data suggests that the fitted model is useful for predicting plant height.

b. $\alpha = 0.05$. From the MINITAB output the t-ratio for b_1 is 6.57, and the t-ratio for b_2 is -7.69. The P-values for the testing $\beta_1 = 0$ and $\beta_2 = 0$ would be twice the area under the 7 df t curve to the right of 6.57 and 7.69, respectively. From Appendix Table 4, the P-values are found to be practically zero. Both hypotheses would be rejected. The data suggests that both the linear and quadratic terms are important.

c. The point estimate of the mean y value when x = 2 is $\hat{y}$ = 41.74 + 6.581(2) – 2.36(4) = 45.46.

The 95% confidence interval is $45.46 \pm (2.37)(1.037) \Rightarrow 45.46 \pm 2.46 \Rightarrow (43.0, 47.92)$. With 95% confidence, the mean height of wheat plants treated with x = 2 (10^2 = 100 uM of Mn) is estimated to be between 43 and 47.92 cm.

d. The point estimate of the mean y value when x = 1 is $\hat{y}$ = 41.74 + 6.58(1) – 2.36(1) = 45.96.

The 90% confidence interval is $45.96 \pm (1.9)(1.031) \Rightarrow 45.96 \pm 1.96 \Rightarrow (44.0, 47.92)$. With 90% confidence, the mean height of wheat plants treated with x = 1 (10 = 10 uM of Mn) is estimated to be between 44 and 47.92 cm.

Exercises 14.49 – 14.59

14.49 One possible way would have been to start with the set of predictor variables consisting of all five variables, along with all quadratic terms, and all interaction terms. Then, use a selection procedure like backward elimination to arrive at the given estimated regression equation.

14.51 The model using the three variables x_3, x_9, x_{10} appears to be a good choice. It has an adjusted R^2 which is only slightly smaller than the largest adjusted R^2. This model is almost as good as the model with the largest adjusted R^2 but has two less predictors.

14.53 **a.** The model has 9 predictors.

H_o: $\beta_1 = \beta_2 = \beta_3 = \beta_4 = \beta_5 = \beta_6 = \beta_7 = \beta_8 = \beta_9 = 0$

H_a: at least one among β_1, β_2, β_3, β_4, β_5, β_6, β_7, β_8, β_9 is not zero

$\alpha = 0.05$ (for illustration).

$F = \dfrac{R^2/k}{(1-R^2)/[n-(k+1)]}$

n = 1856, $df_1 = k = 9$, $df_2 = n - (k + 1) = 1856 - 10 = 1846$.

$R^2 = = 0.3092$.

$F = \dfrac{R^2/k}{(1-R^2)/[n-(k+1)]} = \dfrac{0.3092/9}{(1-0.3092)/1846} = 91.8$

From Appendix Table 6, 0.001 > P-value.

Since the P-value is less than α, the null hypothesis is rejected. There does appear to be a useful linear relationship between ln(blood cadmium level) and at least one of the nine predictors.

b. If a backward elimination procedure was followed in the stepwise regression analysis, then the statements in the paper suggest that all variables except daily cigarette consumption and alcohol consumption were eliminated from the model. Of the two predictors left in the model, cigarette consumption would have a larger t-ratio than alcohol consumption.

There is an alternative procedure called the forward selection procedure which is available in most statistical software packages including MINITAB. According to this method one starts with a model having only the intercept term and enters one predictor at a time into the model. The predictor explaining most of the variance is entered first. The second predictor entered into the model is the one that explains most of the remaining variance, and so on. If the forward selection method was followed in the current problem then the statements in the paper would suggest that the variable to enter the model first is daily cigarette consumption and the next variable to enter the model is alcohol consumption. No further predictors were entered into the model.

14.55 a. Yes, they do show a similar pattern.

b. Standard error for the estimated coefficient of log of sales = (estimated coefficient)/t-ratio = 0.372/6.56 = 0.0567.

c. The predictor with the smallest (in magnitude) associated t-ratio is Return on Equity. Therefore it is the first candidate for elimination from the model. It has a t-ratio equal to 0.33 which is much less than $t_{out} = 2.0$. Therefore the predictor Return on Equity would be eliminated from the model if a backward elimination method is used with $t_{out} = 2.0$.

d. No. For the 1992 regression, the first candidate for elimination when using a backward elimination procedure is CEO Tenure since it has the smallest t-ratio (in magnitude).

e. We test H_o: Coefficient of Stock Ownership is equal to 0 versus H_a : Coefficient of Stock Ownership is less than 0. The t-ratio for this test is -0.58. Using Table 4 from the appendix we find that the area to the left of the 153 d.f. t curve is approximately 0.3. So, the P-value for the test is approximately 0.3. Using MINITAB we find that the exact P-value is 0.2814.

14.57 Using MINITAB, the best model with k variables has been found and summary statistics from each are given below.

Number of Variables	Variables Included	R^2	Adjusted R^2	Cp
1	x_4	0.824	0.819	14.0
2	x_2, x_4	0.872	0.865	2.9
3	x_2, x_3, x_4	0.879	0.868	3.1
4	x_1, x_2, x_3, x_4	0.879	0.865	5.0

The best model, using the procedure of minimizing Cp, would use variables x_2, x_4. Hence, the set of predictor variables selected here is not the same as in problem

14.59 Using MINITAB, the best model with k variables has been found and summary statistics for each are given below.

k	Variables Included	R^2	Adjusted R^2	Cp
1	x_4	0.067	0.026	5.8
2	x_2, x_4	0.111	0.031	6.6
3	x_1, x_3, x_4	0.221	0.110	5.4
4	x_1, x_3, x_4, x_5	0.293	0.151	5.4
5	x_1, x_2, x_3, x_4, x_5	0.340	0.166	6.0

It appears that the model using x_1, x_3, x_4 is the best model, using the criterion of minimizing Cp.

Exercises 14.61 – 14.71

14.61 **a.** $H_o: \beta_1 = \beta_2 = \beta_3 = \ldots = \beta_{11} = 0$

H_a: at least one among β_i's is not zero

$\alpha = 0.01$

$$F = \frac{R^2/k}{(1-R^2)/[n-(k+1)]}$$

$n = 88$, $df_1 = k = 11$, $df_2 = n - (k + 1) = 88 - 12 = 76$

$$F = \frac{R^2/k}{(1-R^2)/[n-(k+1)]} = \frac{.64/11}{(1-.64)/76} = \frac{.058182}{.004737} = 12.28$$

Appendix Table 6 does not have entries for $df_1 = 11$, but using $df_1 = 10$ it can be determined that $0.001 > $ P-value.

Since the P-value is less than α, the null hypothesis is rejected. There does appear to be a useful linear relationship between y and at least one of the predictors.

b. Adjusted $R^2 = 1 - \left[\frac{n-1}{n-(k+1)}\right]\frac{SSResid}{SSTo}$

To calculate adjusted R^2, we need the values for SSResid and SSTo. From the information given, we obtain:

$$s_e^2 = (5.57)^2 \Rightarrow 31.0249 = \frac{SSResid}{88-12} \Rightarrow SSResid = 76(31.0249) = 2357.8924$$

$$R^2 = .64 \Rightarrow .64 = 1 - \frac{SSResid}{SSTo} \Rightarrow .64 = 1 - \frac{2357.8924}{SSTol} \Rightarrow \frac{2357.8924}{SSTo} = .36$$

$$\Rightarrow SSTo = \frac{2357.8924}{.36} = 6549.7011.$$

So, $Adjusted\ R^2 = 1 - \left[\frac{n-1}{n-(k+1)}\right]\frac{SSResid}{SSTo} = 1 - \frac{87}{76}\left(\frac{2357.8924}{6549.7011}\right) = 1 - .4121 = .5879.$

c. t-ratio $= \dfrac{b_1 - 0}{s_{b_1}} = 3.08 \Rightarrow s_{b_1} = \dfrac{b_1}{3.08} = \dfrac{.458}{3.08} = .1487$

$b_1 \pm (t \; critical)s_{b_1} \Rightarrow .458 \pm (2.00)(.1487) \Rightarrow .458 \pm .2974 \Rightarrow (.1606, .7554)$

From this interval, we estimate the value of β_1 to be between –0.1606 and 0.7554.

d. Many of the variables have t-ratios that are close to zero. The one with the smallest t in absolute value is x_9: certificated staff-pupil ratio. For this reason, I would eliminate x_9 first.

e. H_o: $\beta_3 = \beta_4 = \beta_5 = \beta_6 = 0$

H_a: at least one among $\beta_3, \beta_4, \beta_5, \beta_6$ is non-zero.

None of the procedures presented in this chapter could be used. The two procedures presented tested "all variables as a group" or "a single variable's contribution".

14.63 a.

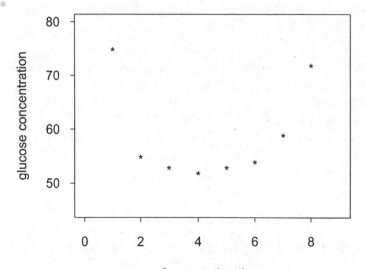

Based on this scatterplot a quadratic model in x is suggested.

b. H_o: $\beta_1 = \beta_2 = 0$

H_a: At least one of the two β_i's is not zero.

$\alpha = 0.05$

Test statistic: $F = \dfrac{SSRegr\,/\,k}{SSResid\,/\,[n - (k+1)]}$

$F = \dfrac{525.11/2}{61.77/5} = 21.25$

From Appendix Table 6, 0.01 > P-value > 0.001.

Since the P-value is less than α, the null hypothesis is rejected. The data suggests that the quadratic model is useful for predicting glucose concentration.

c. H_o: $\beta_2 = 0$ H_a: $\beta_2 \neq 0$

$\alpha = 0.05$

The test statistic is: $t = \dfrac{b_2}{s_{b_2}}$ with df. = 5.

$t = \dfrac{1.7679}{.2712} = 6.52$

P-value = 2(area under the 5 df t curve to the right of 6.52) ≈ 2(0) = 0.

Since the P-value is less than α, the null hypothesis is rejected. The data suggests that the quadratic term cannot be eliminated from the model.

14.65 When n = 21 and k = 10, Adjusted R^2 = 1 – 2(SSResid/SSTo).

Then Adjusted R^2 < 0 ⇒ ½ < SSResid / SSTo = 1 – R^2 ⇒ 1/2 > R^2.

Hence, when n = 21 and k = 10, Adjusted R^2 will be negative for values of R^2 less than 0.5.

14.67 First, the model using all four variables was fit. The variable age at loading (x_3) was deleted because it had the t-ratio closest to zero and it was between –2 and 2. Then, the model using the three variables x_1, x_2, and x_4 was fit. The variable time (x_4) was deleted because its t-ratio was closest to zero and was between –2 and 2. Finally, the model using the two variables x_1 and x_2 was fit. Neither of these variables could be eliminated since their t-ratios were greater than 2 in absolute magnitude. The final model then, includes slab thickness (x_1) and load (x_2). The predicted tensile strength for a slab that is 25 cm thick, 150 days old, and is subjected to a load of 200 kg for 50 days is $\hat{y}$ = 13 – 0.487(25) + 0.0116(200) = 3.145.

14.69 a.

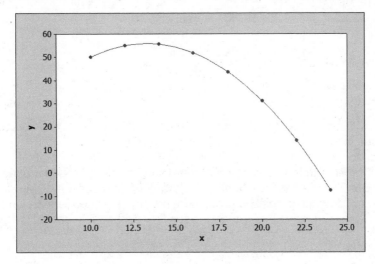

b. The claim is very reasonable because 14 is close to where the smooth curve has its highest value.

14.71 **a.** Output from MINITAB is given below.

The regression equation is: $Y = 1.56 + .0237\ X1 - 0.000249\ X2$.

Predictor	Coef	Stdev	t-ratio	p
Constant	1.56450	0.07940	19.70	0.000
X1	0.23720	0.05556	4.27	0.000
X2	−0.00024908	0.00003205	−7.77	0.000

$s = 0.05330$ R-sq = 86.5%   R-sq(adj) = 85.3%

Analysis of Variance

SOURCE	DF	SS	MS	F	p
Regression	2	0.40151	0.20076	70.66	0.000
Error	22	0.06250	0.00284		
Total	24	0.46402			

b. $H_o: \beta_1 = \beta_2 = 0$
   H_a: At least one of the two β_i's is not zero.
   $\alpha = 0.05$
   Test statistic: $F = \dfrac{SSRegr\ /\ k}{SSResid\ /\ [n - (k+1)]}$

   $F = \dfrac{.40151/2}{.0625/22} = 70.67$

From the Minitab output, the P-value associated with the F test is practically zero. Since the P-value is less than α, the null hypothesis is rejected.

c. The value for R^2 is 0.865. This means that 86.5% of the total variation in the observed values for profit margin has been explained by the fitted regression equation. The value for s_e is 0.0533. This means that the typical deviation of an observed value from the predicted value is 0.0533, when predicting profit margin using this fitted regression equation.

d. No. Both variables have associated t-ratios that exceed 2 in absolute magnitude. Hence, neither can be eliminated from the model.

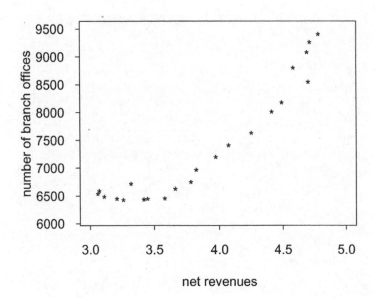

There do not appear to be any influential observations. However, there is substantial evidence of multicollinearity. The plot shows a pronounced linear relationship between x_1 and x_2. This is evidence of multicollinearity between x_1 and x_2.

Chapter 15

15.1   **a.** $0.001 < $ P-value < 0.01
      b. P-value > 0.10
      c. P-value $= 0.01$
      d. $0.001 > $ P-value
      e. $0.05 < $ P-value < 0.10
      f. $0.01 < $ P-value < 0.05 (Using $df_2 = 40$ and $df_2 = 60$ tables).

15.3   **a.** Let μ_1, μ_2, μ_3 and μ_4 denote the true average length of stay in a hospital for health plans 1, 2, 3 and 4 respectively.
      H_o: $\mu_1 = \mu_2 = \mu_3 = \mu_4$
      H_a: At least two of the four μ_i's are different.
    b. $df_1 = 4 - 1 = 3$ $df_2 = 32 - 4 = 28$ $\alpha = 0.01$
    From Appendix Table 6, $0.05 > $ P-value > 0.01,
    Since the P-value exceeds α, H_o is not rejected. Hence, it would be concluded that the average length of stay in the hospital is the same for the four health plans.
    c. $df_1 = 4 - 1 = 3$ $df_2 = 32 - 4 = 28$ $\alpha = 0.01$
    From Appendix Table 6, $0.05 > $ P-value > 0.01. Since the P-value exceeds α, H_o is not rejected. Therefore, the conclusion would be the same.

15.5   Let μ_i denote the mean compression strength for each of the four types of boxes i (i = 1, 2, 3, 4,)

$$H_0 : \mu_1 = \mu_2 = \mu_3 = \mu_4$$

H_a : At least two of the four μ_i's are different.

$\alpha = 0.01$

Test statistic: $F = \dfrac{\text{MSTr}}{\text{MSE}}$

The boxplots below show the four samples are slightly skewed but with no outliers. The largest standard deviation ($s_1 = 46.55$) is not more than twice the smallest standard deviation ($s_3 = 37.20$). The data consists of independently chosen random samples. The assumptions of ANOVA are reasonable.

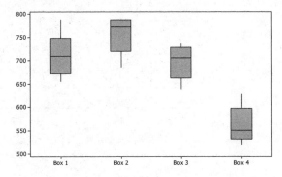

229

N = 4(6) = 24

$$SSTr = n_1(\bar{x}_1 - \bar{\bar{x}})^2 + n_2(\bar{x}_2 - \bar{\bar{x}})^2 + ... + n_k(\bar{x}_k - \bar{\bar{x}})^2$$

$$= 6(713 - 682.5)^2 + 6(756.93 - 682.5)^2 + 6(698.07 - 682.5)^2 + 6(562.02 - 682.5)^2$$

$$= 127367.58$$

treatment df = k − 1 = 3 $MSTr = \dfrac{127367.58}{3} = 42455.86$

$$SSE = (n_1 - 1)s_1^2 + (n_2 - 1)s_2^2 + (n_3 - 1)s_3^2 + (n_4 - 1)s_4^2$$

$$= 5(2166.9) + 5(1627.3) + 5(1383.8) + 5(1589.6)$$

$$= 33838$$

error df = N - k = 24 − 4 = 20. $MSE = \dfrac{33838}{20} = 1691.9$

$$F = \frac{MSTr}{MSE} = \frac{42455.86}{1691.9} = 25.09$$

From Appendix Table 6, P-value < 0.001.
Since the P-value is smaller than α, the null hypothesis is rejected. The data strongly supports the claim that true mean compression strength is not the same for all four box types.

15.7   **a.** The required boxplot obtained using MINITAB is shown below. Price per acre values appear to be similar for 1996 and 1997 but 1998 values are higher. The mean price per acre values for each year are also plotted as a solid square within each box plot.

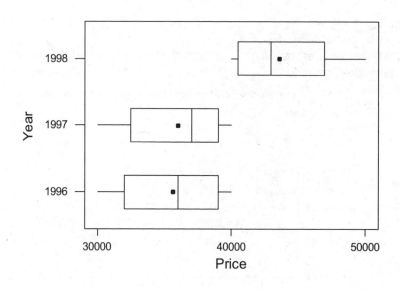

b. Let μ_i denote the mean price per acre for vineyards in year i (i = 1, 2, 3).

H_o: $\mu_1 = \mu_2 = \mu_3$
H_a: At least two of the three μ_i's are different.
$\alpha = 0.01$

Test statistic: $F = \dfrac{MSTr}{MSE}$

$df_1 = k - 1 = 2 \quad df_2 = N - k = 15\text{-}3 = 12.$
$\bar{x}_1 = 35600, \bar{x}_2 = 36000, \bar{x}_3 = 43600$

$$\bar{\bar{x}} = \frac{[5(35600) + 5(36000) + 5(43600)]}{(15)} = 38400$$

MSTr = $[5(35600 - 38400)^2 + 5(36000 - 38400)^2 + 5(43600 - 38400)^2]/2 = $
101600000
$s_1 = 3847.077, s_2 = 3807.887, s_3 = 3911.521$
MSE = $[(5-1)(3847.077)^2 + (5-1)(3807.887)^2 + (5-1)(3911.521)^2]/12 = 14866667$

$$F = \frac{MSTr}{MSE} = \frac{101600000}{14866667} = 6.83$$

From Appendix Table 6, 0.05 > P-value > 0.01.

Since the P-value exceeds α, the null hypothesis is not rejected. At a significance level of $\alpha = 0.01$, the data does not support the claim that the true mean price per acre for the three years under consideration are different.

15.9 Let μ_i denote the mean level of chlorophyll concentration for plants in variety i (i = 1, 2, 3, 4).

H_o: $\mu_1 = \mu_2 = \mu_3 = \mu_4$
H_a: At least two of the four μ_i's are different.
$\alpha = 0.05$

Test statistic: $F = \dfrac{MSTr}{MSE}$

$df_1 = k - 1 = 3 \quad df_2 = N - k = 16.$

$$\bar{\bar{x}} = \frac{[5(.3) + 5(.24) + 4(.41) + 6(.33)]}{(20)} = 0.316$$

MSTr = $[5(0.3 - 0.316)^2 + 5(0.24 - 0.316)^2 + 4(0.41 - 0.316)^2 + 6(0.33 - 0.316)^2]/3 = $
0.06668/3 = 0.022227

$$F = \frac{MSTr}{MSE} = \frac{.022227}{.013} = 1.71$$

From Appendix Table 6, P-value > 0.10.

Since the P-value exceeds α, the null hypothesis is not rejected. The data does not suggest that true mean chlorophyll concentration differs for the four varieties.

15.11 Let μ_i denote the mean Hopkins score for the three groups i (i = 1, 2, 3), where 1 = Soccer athletes, 2 = Non-soccer athletes and, 3 = Control.

$$H_0 : \mu_1 = \mu_2 = \mu_3$$

H_a : At least two of the three μ_i's are different.

$\alpha = 0.05$

Test statistic: $F = \dfrac{MSTr}{MSE}$

We don't have the raw data so but the sample sizes are large so we can assume normality in the sampling distribution. The largest standard deviation ($s_2 = 5.14$) is not more than twice the smallest standard deviation ($s_1 = 3.73$). The data consists of random samples from the three populations of interest. The assumptions of ANOVA are reasonable.

N = 86 + 95 + 53 = 234

$$SSTr = n_1(\bar{x}_1 - \bar{\bar{x}})^2 + n_2(\bar{x}_2 - \bar{\bar{x}})^2 + ... + n_k(\bar{x}_k - \bar{\bar{x}})^2$$

$$= 86(29.9 - 30.19)^2 + 95(30.94 - 30.19)^2 + 53(29.32 - 30.19)^2$$

$$= 100.8$$

treatment df = k − 1 = 2 $MSTr = \dfrac{100.8}{2} = 50.4$

$$SSE = (n_1 - 1)s_1^2 + (n_2 - 1)s_2^2 + (n_3 - 1)s_3^2 + (n_4 - 1)s_4^2$$

$$= 85(13.9) + 94(26.4) + 52(14.3)$$

$$= 4406.7$$

error df = N - k = 234 − 3 = 231.   MSE = $\dfrac{4406.7}{231} = 19.08$

$F = \dfrac{MSTr}{MSE} = \dfrac{50.04}{19.08} = 2.62$

From Appendix Table 6, 0.05 < P-value < 0.1.

Since the P-value is greater than α, the null hypothesis cannot be rejected. The data does not provide sufficient evidence to conclude that true mean Hopkins score is not the same for the three student populations.

15.13 Let μ_i denote the mean dry weight for concentration level i (i = 1, 2, ..., 10).

H_0: $\mu_1 = \mu_2 = ... = \mu_{10}$

H_a: At least two of the ten μ_i's are different.

$\alpha = 0.05$

Test statistic: $F = \dfrac{MSTr}{MSE}$

$df_1 = k − 1 = 9$ $df_2 = N − k = 30$.

$F = \dfrac{MSTr}{MSE} = 1.895$

From Appendix Table 6, 0.10 > P-value > 0.05.

Since the P-value exceeds α, the null hypothesis is not rejected. The data are consistent with the hypothesis that the true mean dry weight does not depend on the level of concentration.

15.15

Source of Variation	Degrees of Freedom	Sum of Squares	Mean Square	F
Treatments	3	75081.72	25027.24	1.70
Error	16	235419.04	14713.69	
Total	19	310500.76		

Let μ_i denote the mean number of miles to failure for brand i sparkplugs (i = 1, 2, 3, 4).
H_0: $\mu_1 = \mu_2 = \mu_3 = \mu_4$
H_a: At least two of the four μ_i's are different.
$\alpha = 0.05$

Test statistic: $F = \dfrac{MSTr}{MSE}$

$df_1 = k - 1 = 3$ $df_2 = N - k = 16$.   From the ANOVA table, F = 1.70.
From Appendix Table 6, P-value > 0.10.
Since the P-value exceeds α, the null hypothesis is not rejected. The data are consistent with the hypothesis that there is no difference between the mean number of miles to failure for the four brands of sparkplugs.

15.17 Computations: $\bar{\bar{x}} = [96(2.15) + 34(2.21) + 86(1.47) + 206(1.69)]/422$
$= 756.1/422 = 1.792$

MSTr = $[96(2.15 - 1.792)^2 + 34(2.21 - 1.792)^2 + 86(1.47 - 1.792)^2$
$+ 206(1.69 - 1.792)^2]/3 = 29.304/3 = 9.768$

$MSE = \dfrac{MSTr}{F} = \dfrac{9.768}{2.56} = 3.816$

Source of Variation	Degrees of Freedom	Sum of Squares	Mean Square	F
Treatments	3	29.304	9.768	2.56
Error	418	1595.088	3.816	
Total	421	1624.392		

Let μ_i denote the mean number of hours per month absent for employees of group i (i = 1, 2, 3, 4).
H_0: $\mu_1 = \mu_2 = \mu_3 = \mu_4$
H_a: At least two of the four μ_i's are different.

$\alpha = 0.01$

Test statistic: $F = \dfrac{MSTr}{MSE}$

$df_1 = k - 1 = 3$ $df_2 = N - k = 418$. From the ANOVA table, $F = 2.56$.
From Appendix Table 6, $0.10 > \text{P-value} > 0.05$.
Since the P-value exceeds α, the null hypothesis is not rejected. The data are consistent with the hypothesis that there is no difference between the mean number of hours per month absent for employees in the four groups.

15.19 Let μ_1, μ_2, and μ_3 denote the true mean fog indices for *Scientific America*, *Fortune*, and *New Yorker*, respectively.

H_o: $\mu_1 = \mu_2 = \mu_3$
H_a: At least two of the three μ_i's are different.
$\alpha = 0.01$

Test statistic: $F = \dfrac{MSTr}{MSE}$

$df_1 = k - 1 = 2$ $df_2 = N - k = 15$.
Computations: $\bar{\bar{x}} = 9.666$

Magazine	Mean	Standard Deviation
S.A.	10.968	2.647
F.	10.68	1.202
N.Y.	7.35	1.412

$MSTr = [6(10.968 - 9.666)^2 + 6(10.68 - 9.666)^2 + 6(7.35 - 9.666)^2]/2 = 48.524/2 = 24.262$
$MSE = [(2.647)^2 + (1.202)^2 + (1.412)^2]/3 = 10.443/3 = 3.4812$

Source of Variation	Degrees of Freedom	Sum of Squares	Mean Square	F
Treatments	2	48.524	24.2620	6.97
Error	15	55.218	3.4812	
Total	17	100.742		

$F = \dfrac{MSTr}{MSE} = \dfrac{24.262}{3.4812} = 6.97$

From Appendix Table 6, $0.01 > \text{P-value} > 0.001$.
Since the P-value is less than α, the null hypothesis is rejected. The data suggests that there is a difference between at least two of the mean fog index levels for advertisements appearing in the three magazines.

15.21 **a.** From the given information:

$n_1 = 6$, $\bar{x}_1 = 10.125$, $s_1^2 = 2.093750$ $\qquad$ $n_2 = 6$, $\bar{x}_2 = 11.375$, $s_2^2 = 3.59375$

$n_3 = 6$, $\bar{x}_3 = 11.708333$, $s_3^2 = 2.310417$ $\qquad$ $n_4 = 5$, $\bar{x}_4 = 12.35$, $s_4^2 = .925$

$$\bar{\bar{x}} = \frac{T}{N} = \frac{261}{23} = 11.347826$$

SSTr $= 6(10.125 - 11.347826)^2 + 6(11.375 - 11.347826)^2 + 6(11.708333 - 11.347826)^2 + 5(12.35 - 11.347826)^2$

$\qquad = 8.971822 + 0.004431 + 0.779791 + 5.021763 = 14.7778$

SSE $= 5(2.093750) + 5(3.59375) + 5(2.310417) + 4(0.925) = 43.689585$.

$df_1 = k - 1 = 4 - 1 = 3$ $\quad$ $df_2 = (6 + 6 + 6 + 5) - 4 = 19$

$$MSTr = \frac{SSTr}{k-1} = \frac{14.7778}{3} = 4.925933$$

$$MSE = \frac{SSE}{N-k} = \frac{43.689585}{19} = 2.299452$$

$$F = \frac{MSTr}{MSE} = \frac{4.925933}{2.299452} = 2.142$$

From Appendix Table 6, P-value > 0.10.

b. $H_0: \mu_1 = \mu_2 = \mu_3 = \mu_4$

H_a: At least two of the four μ_i's are different.

$\alpha = 0.05$

Test statistic: $F = \dfrac{MSTr}{MSE} = 2.142$

$df_1 = k - 1 = 3$ $\quad$ $df_2 = N - k = 19$. From the ANOVA table, $F = 2.142$.

From Appendix Table 6, P-value > 0.10.

Since the P-value exceeds α, the null hypothesis is not rejected. The data does not suggest that there are differences in true average responses among the treatments.

Exercises 15.23 – 15.33

15.23 Since the intervals for $\mu_1 - \mu_2$ and $\mu_1 - \mu_3$ do not contain zero, μ_1 and μ_2 are judged to be different and μ_1 and μ_3 are judged to be different. Since the interval for $\mu_2 - \mu_3$ contains zero, μ_2 and μ_3 are judged not to be different. Hence, statement (iii) best describes the relationship between μ_1, μ_2, and μ_3.

15.25 $k = 4$ $\qquad$ Error df $= 4(20) - 4 = 80 - 4 = 76$

Appendix Table 7, gives the 99% Studentized range critical value $q = 4..59$ (using df = 60, as the closest value for df = 76). As the sample size is the same for each group, the equation can be simplified:

$\mu_1 - \mu_2$: $(16.30 - 15.25) \pm 4.59\sqrt{\left(\dfrac{7.7861}{20}\right)}$ $\Rightarrow$ $1.05 \pm 2.86 \Rightarrow (-1.81, 3.91)$ ← *Includes 0*

$\mu_1 - \mu_3$: $(16.30 - 12.05) \pm 4.59\sqrt{\left(\dfrac{7.7861}{20}\right)}$ $\Rightarrow$ $4.25 \pm 2.86 \Rightarrow (1.39, 7.11)$ ← *Does not include 0*

$\mu_1 - \mu_4$: $(16.30 - 9.30) \pm 4.59\sqrt{\left(\dfrac{7.7861}{20}\right)}$ $\Rightarrow$ $7 \pm 2.86 \Rightarrow (4.14, 9.88)$ ← *Does not include 0*

$\mu_2 - \mu_3$: $(15.25 - 12.05) \pm 4.59\sqrt{\left(\dfrac{7.7861}{20}\right)}$ $\Rightarrow$ $3.2 \pm 2.86 \Rightarrow (0.34, 6.06)$ ← *Does not include 0*

$\mu_2 - \mu_4$: $(15.25 - 9.30) \pm 4.59\sqrt{\left(\dfrac{7.7861}{20}\right)}$ $\Rightarrow$ $5.95 \pm 2.86 \Rightarrow (3.09, 8.81)$ ← *Does not include 0*

$\mu_3 - \mu_4$: $(12.05 - 9.30) \pm 4.59\sqrt{\left(\dfrac{7.7861}{20}\right)}$ $\Rightarrow$ $2.75 \pm 2.86 \Rightarrow (-0.11, 5.61)$ ← *Includes 0*

μ_1 differs from μ_3; μ_1 differs from μ_4; μ_2 differs from μ_3; μ_2 differs from μ_4.

Method	4	3		2	1
$\overline{x}$	9.30	12.05		15.25	16.30

15.27 $k = 4$ Error df $= (5 + 5 + 4 + 6) - 4 = 16$
From Appendix Table 7, q = 4.05 for 95% confidence.

$\mu_1 - \mu_2 : (.30 - .24) \pm 4.05\sqrt{\dfrac{.013}{2}\left(\dfrac{1}{5} + \dfrac{1}{5}\right)}$ $\Rightarrow .06 \pm .2065 \Rightarrow (-.1465, .2665)$

$\mu_1 - \mu_3 : (.30 - .41) \pm 4.05\sqrt{\dfrac{.013}{2}\left(\dfrac{1}{5} + \dfrac{1}{4}\right)}$ $\Rightarrow -.11 \pm .2190 \Rightarrow (-.3190, .1090)$

$\mu_1 - \mu_4 : (.30 - .33) \pm 4.05\sqrt{\dfrac{.013}{2}\left(\dfrac{1}{5} + \dfrac{1}{6}\right)}$ $\Rightarrow -.03 \pm .1977 \Rightarrow (-.2277, .1677)$

$\mu_2 - \mu_3 : (.24 - .41) \pm 4.05\sqrt{\dfrac{.013}{2}\left(\dfrac{1}{5} + \dfrac{1}{4}\right)}$ $\Rightarrow -.17 \pm .2190 \Rightarrow (-.3890, .0490)$

$\mu_2 - \mu_4 : (.24 - .33) \pm 4.05\sqrt{\dfrac{.013}{2}\left(\dfrac{1}{5} + \dfrac{1}{6}\right)}$ $\Rightarrow -.09 \pm .1977 \Rightarrow (-.2877, .1077)$

$\mu_3 - \mu_4 : (.41 - .33) \pm 4.05\sqrt{\dfrac{.013}{2}\left(\dfrac{1}{4} + \dfrac{1}{6}\right)}$ $\Rightarrow .08 \pm .2108 \Rightarrow (-.1308, .2908)$

There are no pairwise differences.

Variety	2. RO	1. BI	4. TO	3. WA
Mean	0.24	0.30	0.33	0.41

No differences are detected. This conclusion is in agreement with the results of the F test done in problem 15.9.

15.29

Group	Simultaneous	Sequential	Control
Mean	$\bar{x}_3$	$\bar{x}_2$	$\bar{x}_1$

15.31 The mean water loss when exposed to 4 hours fumigation is different from all other means. The mean water loss when exposed to 2 hours fumigation is different from that for levels 16 and 0, but not 8. The mean water losses for duration 16, 0, and 8 hours are not different from one another. No other differences are significant.

15.33 $k = 3$ Error df $= (24 + 24 + 20) - 3 = 65$

From Appendix Table 7, the value of q is between 3.40 and 3.36. The value of 3.40 will be used in the computations.

$$\mu_1 - \mu_2 : (6.60 - 5.37) \pm 3.40 \sqrt{\frac{2.028}{2}\left(\frac{1}{24} + \frac{1}{24}\right)} \Rightarrow 1.23 \pm 3.40(.291) \Rightarrow 1.23 \pm .99 \Rightarrow (.24, 2.22)$$

$$\mu_1 - \mu_3 : (6.60 - 5.20) \pm 3.40 \sqrt{\frac{2.028}{2}\left(\frac{1}{24} + \frac{1}{20}\right)} \Rightarrow 1.40 \pm 3.40(.305) \Rightarrow 1.40 \pm 1.04 \Rightarrow (.36, 2.44)$$

$$\mu_2 - \mu_3 : (5.37 - 5.20) \pm 3.40 \sqrt{\frac{2.028}{2}\left(\frac{1}{24} + \frac{1}{20}\right)} \Rightarrow .17 \pm 3.40(.305) \Rightarrow .17 \pm 1.04 \Rightarrow (-.87, 1.21)$$

So μ_1 differs from μ_2 and μ_3, but μ_2 and μ_3 do not differ.

Schedule	3	2	1
$\bar{x}$	5.20	5.37	6.60

Exercises 15.35 – 15.43

15.35 A randomized block experiment was used to control the factor <u>value of house</u>, which definitely affects the assessors' appraisals. If a completely randomized experiment had been done, then there would have been danger of having the assessors appraising houses which were not of similar values. Therefore, differences between assessors would be partly due to the fact that the homes were dissimilar, as well as to differences in the appraisals made.

15.37 a.

Source of Variation	Degrees of Freedom	Sum of Squares	Mean Square	F
Treatments	2	11.7	5.85	0.37
Blocks	4	113.5	28.375	
Error	8	125.6	15.7	
Total	14	250.8		

b. H_0: The mean appraised value does not depend on which assessor is doing the appraisal.
H_a: The mean appraised value does depend on which assessor is doing the appraisal.
$\alpha = 0.05$

Test statistic: $F = \dfrac{MSTr}{MSE}$

$df_1 = k - 1 = 2$ $df_2 = (k - 1)(l - 1) = 8$. From the ANOVA table, F = 0.37.
From Appendix Table 6, P-value > 0.10.

Since the P-value exceeds α, the null hypothesis is not rejected. The mean appraised value does not seem to depend on which assessor is doing the appraisal.

15.39 a. Other environmental factors (amount of rainfall, number of days of cloudy weather, average daily temperature, etc.) vary from year to year. Using a randomized complete block helps control for variation in these other factors.

b. $SSTr = 3[(138.33 - 149)^2 + (152.33 - 149)^2 + (156.33 - 149)^2] = 536.00$
$SSBl = 3[(173 - 149)^2 + (123.67 - 149)^2 + (150.33 - 149)^2] = 3658.13$

Source of Variation	Degrees of Freedom	Sum of Squares	Mean Square	F
Treatments	2	536.00	268.00	14.91
Blocks	2	3658.13	1829.07	
Error	4	71.87	17.97	
Total	8	4266.00		

H_o: The mean height does not depend on the rate of application of effluent.
H_a: The mean height does depend on the rate of application of effluent.
$\alpha = 0.05$

Test statistic: $F = \dfrac{MSTr}{MSE}$

$df_1 = k - 1 = 2$ and $df_2 = (k - 1)(l - 1) = 4$. From the ANOVA table, $F = 14.91$.
From Appendix Table 6, $0.05 > $ P-value > 0.01.

Since the P-value is less than α, the null hypothesis is rejected. The data suggests that the mean height of cotton plants does depend on the rate of application of effluent.

c. Since the sample sizes are equal, the $\pm$ factor is the same for each comparison.
$k = 3$ Error df $= 4$ $q = 5.04$ (This value came from a more extensive table of critical values for the Studentized range distribution than the text provides.)

The $\pm$ factor is: $5.04\sqrt{\dfrac{17.97}{3}} = 12.33$

Application Rate	1	2	3
Mean	138.33	152.33	156.33

The mean height under an application rate of 350 differs from those using application rates of 440 and 515.

The mean height using application rates of 440 and 515 do not differ.

15.41 H_o: The mean height does not depend on the seed source.
H_a: The mean height does depend on the seed source.
$\alpha = 0.05$

Test statistic: $F = \dfrac{MSTr}{MSE}$

Source of Variation	Degrees of Freedom	Sum of Squares	Mean Square	F
Treatments	4	4.543	1.136	0.86
Blocks	3	7.862	2.621	
Error	12	15.701	1.308	
Total	19	28.106		

$df_1 = k - 1 = 4$, $df_2 = (k - 1)(l - 1) = 12$, and $F = 0.868$.
From Appendix Table 6, P-value > 0.10.

Since the P-value exceeds α, the null hypothesis is not rejected. The data are consistent with the hypothesis that mean height does not depend on the seed source.

15.43

Source of Variation	Degrees of Freedom	Sum of Squares	Mean Square	F
Treatments	2	3.97	1.985	79.4
Blocks	7	0.2503	0.0358	
Error	14	0.3497	0.025	
Total	23	4.57		

H_0: The mean energy use does not depend on the type of oven.
H_a: The mean energy use does depend on the type of oven.
$\alpha = 0.01$

Test statistic: $F = \dfrac{MSTr}{MSE}$

$df_1 = k - 1 = 2$ and $df_2 = (k - 1)(l - 1) = 14$. From the ANOVA table, $F = 79.4$.
From Appendix Table 6, $0.001 >$ P-value.
Since the P-value is less than α, the null hypothesis is rejected. The data suggests quite strongly that the mean energy use depends on the type of oven used.

Exercises 15.45 – 15.55

15.45   a.

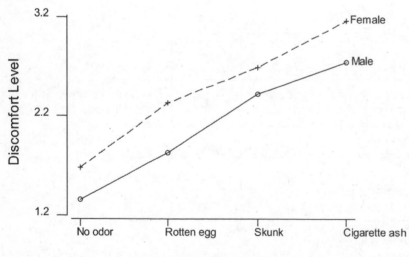

b. The graphs for males and females are very nearly parallel. There does not appear to be an interaction between gender and type of odor.

240

15.47 **a.** The plot does suggest an interaction between peer group and self-esteem. The change in average response, when changing from low to high peer group, is not the same for the low self-esteem group and the high self-esteem group. This is indicated by the non-parallel lines.

b. The change in the average response is greater for the low self-esteem group than it is for the high self-esteem group, when changing from low to high peer group interaction. Therefore, the authors are correct in their statement.

15.49

Source of Variation	Degrees of Freedom	Sum of Squares	Mean Square	F
Size (A)	2	0.088	0.044	4.00
Species (B)	1	0.048	0.048	4.363
Size by Species	2	0.048	0.024	2.18
Error	12	0.132	0.011	
Total	17	0.316		

H_o: There is no interaction between Size (A) and Species (B).
H_a: There is interaction between Size and Species.
$\alpha = 0.01$

The test statistic is: $F_{AB} = \dfrac{MSAB}{MSE}$.

$df_1 = 2$, $df_2 = 12$, and $F_{AB} = 2.18$
From Appendix Table 6, P-value > 0.10.
Since the P-value exceeds α, the null hypothesis is not rejected. The data are consistent with the hypothesis of no interaction between Size and Species. Hence, hypothesis tests on main effects will be done.

H_o: There are no size main effects.
H_a: There are size main effects.
$\alpha = 0.01$

The test statistic is: $F_A = \dfrac{MSA}{MSE}$.

$df_1 = 2$, $df_2 = 12$, and $F_A = 4.00$.
From Appendix Table 6, 0.05 > P-value > 0.01.
Since the P-value exceeds α, the null hypothesis of no size main effects is not rejected. When using $\alpha=0.01$, the data support the conclusion that there are no differences between the mean preference indices for the three sizes of bass.

H_o: There are no species main effects.
H_a: There are species main effects.
$\alpha = 0.01$

The test statistic is: $F_B = \dfrac{MSB}{MSE}$.

$df_1 = 1$, $df_2 = 12.$, and $F_B = 4.363$
From Appendix Table 6, P-value > 0.05.
Since the P-value exceeds α, the null hypothesis of no species main effects is not rejected. At a significance level of $\alpha = 0.01$, the data are consistent with the hypothesis that there are no differences between the mean preference indices for the three species of bass.

15.51

Source of Variation	Degrees of Freedom	Sum of Squares	Mean Square	F
A main effects	2	4206	2103	8.67
B main effects	1	1782	1782.0	7.35
AB interaction	2	1947	973.5	4.01
Error	12	2911	242.58	
Total	17	10846		

H_o: There is no interaction between device (A) and food supply (B).
H_a: There is interaction between device and food supply.
$\alpha = 0.05$

The test statistic is: $F_{AB} = \dfrac{MSAB}{MSE}$.

$df_1 = 2$, $df_2 = 12$, $F_{AB} = 4.01$.
From Appendix Table 6, $0.05 > $ P-value $ > 0.01$.
Since the P-value is less than α, the null hypothesis is rejected. The data suggests that there is interaction between device and food supply. Hence, hypothesis tests on main effects are not appropriate.

15.53 **a.** There are two age classes.
b. There were twenty-one observations made for each age-sex combination.

c.

Source of Variation	Degrees of Freedom	Sum of Squares	Mean Square	F
Age (A)	1	0.614	0.6140	8.73
Sex (B)	1	1.754	1.7540	24.95
Age by Sex (AB)	1	0.146	0.1460	2.08
Error	80	5.624	0.0703	
Total	83	8.138		

H_o: There is no interaction between age (A) and sex (B).
H_a: There is interaction between age and sex.
$\alpha = 0.01$ (for illustration)

Test statistic: $F_{AB} = \dfrac{MSAB}{MSE}$

$df_1 = 1$, $df_2 = 80$, and $F_{AB} = 2.08$. From Appendix Table 6, P-value > 0.10.
Since the P-value exceeds α, the null hypothesis is not rejected. The data are consistent with the hypothesis of no interaction between age and sex. Hence, hypothesis tests on main effects will be done.

H_o: There are no age main effects.
H_a: There are age main effects.
$\alpha = 0.01$

Test statistic: $F_A = \dfrac{MSA}{MSE}$

$df_1 = 1$, $df_2 = 80$, and $F_A = 8.73$. From Appendix Table 6, 0.01 > P-value > 0.001.
Since the P-value is less than α, the null hypothesis of no age main effects is rejected. The data supports the conclusion that there is a difference between the mean territory size for the two age groups.

H_o: There are no sex main effects.
H_a: There are sex main effects.
$\alpha = 0.01$

Test statistic: $F_B = \dfrac{MSB}{MSE}$

$df_1 = 1$, $df_2 = 80$, and $F_B = 24.95$. From Appendix Table 6, 0.001 > P-value.
Since the P-value is less than α, the null hypothesis of no sex main effects is rejected. The data supports the conclusion that there is a difference between the mean territory size for the two sexes.

15.55

Source of Variation	Degrees of Freedom	Sum of Squares	Mean Square	F
Diet	2	18138	9069.0	28.92
Temperature	2	5182	2591.0	8.26
Interaction	4	1737	434.25	1.38
Error	36	11291	313.64	
Total	44	36348		

To test for no interaction: $\alpha = 0.05$
$df_1 = 4$, $df_2 = 36$, and $F_{AB} = 1.38$. From Appendix Table 6, P-value > 0.10.
Since the P-value exceeds α, the null hypothesis of no interaction is not rejected. Thus, tests for main effects are appropriate.
To test for diet main effects: $\alpha = 0.05$
$df_1 = 2$, $df_2 = 36$, and $F_A = 28.92$. From Appendix Table 6, $0.001 >$ P-value.
Since the P-value is less than α, the null hypothesis of no diet main effects is rejected. The data suggests that mean gross daily energy intake depends upon the diet given to the animal.
To test for temperature main effects: $\alpha = 0.05$
$df_1 = 2$, $df_2 = 36$, and $F_B = 8.26$. From Appendix Table 6, $0.001 >$ P-value.
Since the P-value is less than α, the null hypothesis of no temperature main effects is rejected. The data suggests that mean gross daily energy intake depends upon the level of temperature.

Exercises 15.57 – 15.67

15.57 H_o: $\mu_1 = \mu_2 = \mu_3$
H_a: at least two among μ_1, μ_2, μ_3 are not equal.

$$F = \frac{MSTr}{MSE}$$

$N = 90$, $df_1 = 2$ $df_2 = 87$
$T = 30(9.40 + 11.63 + 11.00) = 960.9$

$$\bar{\bar{x}} = \frac{T}{N} = \frac{960.9}{90} = 10.677$$

$SSTr = 30[(9.40 - 10.677)^2 + (11.63 - 10.677)^2 + (11.00 - 10.677)^2]$
       $= 30[1.630729 + 0.908209 + 0.104329] = 79.298$

$$MSTr = \frac{79.298}{2} = 39.649$$

$$MSE = \frac{749.85}{87} = 8.619$$

$$F = \frac{MSTr}{MSE} = \frac{39.649}{8.619} = 4.60$$

From Appendix Table 6, $0.05 > $ P-value > 0.01.
If $\alpha = 0.05$ is used, then the P-value is less than α and the null hypothesis would be rejected. The conclusion would be that the true population mean score is not the same for the three types of students.
If $\alpha = 0.01$ is used, then the P-value exceeds α and the null hypothesis would not be rejected. The conclusion would be that the data are consistent with the hypothesis that the mean score is the same for the three types of students.

15.59

Source of Variation	Degrees of Freedom	Sum of Squares	Mean Square	F
Treatments	4	14.962	3.741	36.7
Blocks	8	9.696	1.212	
Error	32	3.262	0.102	
Total	44	27.920		

H_o: The true mean smoothness score does not differ for the five drying methods.
H_a: The true mean smoothness score differs for at least two of the drying methods.
$\alpha = 0.05$

Test statistic: $F = \dfrac{MSTr}{MSE}$

$df_1 = k - 1 = 4$, $df_2 = (k - 1)(l - 1) = 32$, and from the ANOVA table, $F = 36.7$.
From Appendix Table 6, $0.001 > $ P-value.
Since the P-value is less than α, the null hypothesis is rejected. The true mean smoothness scores differ for at least two of the drying methods.

15.61 Let μ_i denote the mean fill weight of pocket i (i = 1, 2, 3, 4, and 5)
H_o: $\mu_1 = \mu_2 = \mu_3 = \mu_4 = \mu_5$
H_a: at least two among $\mu_1, \mu_2, \mu_3, \mu_4, \mu_5$ are not equal.
$\alpha = 0.05$

Test statistic: $F = \dfrac{MSTr}{MSE}$

Computations: $\bar{\bar{x}} = 9.92, \Sigma x^2 = 2461, \bar{x}_1 = 10.08, \bar{x}_2 = 9.98,$
$\bar{x}_3 = 9.88, \bar{x}_4 = 9.78, \bar{x}_5 = 9.88$
SSTo $= 2461 - 25(9.92)^2 = 0.84$
SSTr $= 5[(10.08)^2 + (9.98)^2 + (9.88)^2 + (9.78)^2 + (9.88)^2] - 25(0.92)^2 = 0.26$

Source of Variation	Degrees of Freedom	Sum of Squares	Mean Square	F
Treatments	4	0.26	0.065	2.24
Error	20	0.58	0.029	
Total	24	0.84		

$df_1 = 4$, $df_2 = 20$, and $F = 2.24$.
From Appendix Table 6, P-value > 0.10.
Since the P-value exceeds α, the null hypothesis is not rejected. The data are consistent with the hypothesis that the mean fill weight is the same for each pocket.

15.63

Source of Variation	Degrees of Freedom	Sum of Squares	Mean Square	F
Locations	14	0.6	0.04286	1.89
Months	11	2.3	0.20909	9.20
Error	154	3.5	0.02273	
Total	179	6.4		

Test for locations effects: $\alpha = 0.05$
$df_1 = 14$ and $df_2 = 154$. The F ratio to test for locations is $F = 1.89$. From Appendix Table 6, $0.05 > $ P-value $ > 0.01$. Since the P-value is less than α, the null hypothesis of no location main effects is rejected. The data suggests that the true concentration differs by location.
Test for months effects: $\alpha = 0.05$
$df_1 = 11$ and $df_2 = 154$. The F ratio to test for months is $F = 9.20$. From Appendix Table 6, $0.001 > $ P-value. Since the P-value is less than α, the null hypothesis of no month main effects is rejected. The data suggests that the true mean concentration differs by month of year.

15.65 Multiplying each observation in a single-factor ANOVA will change $\bar{x}_i$, $\bar{\bar{x}}$, and s_i by a factor of c. Hence, MSTr and MSE will be also changed, but by a factor of c^2. However, the F ratio remains unchanged because $c^2 MSTr / c^2 MSE = MSTr/MSE$. That is, the c^2 in the numerator and denominator cancel. It is reasonable to expect a test statistic not to depend on the unit of measurement.

15.67 $c_1 = 1, c_2 = -.5, c_3 = -.5$

$c_1 \bar{x}_1 + c_2 \bar{x}_2 + c_3 \bar{x}_3 = 44.571 - .5(45.857) - .5(48) = -2.3575$

$$MSE\left[\frac{c_1^2}{n_1} + \frac{c_2^2}{n_2} + \frac{c_3^2}{n_3}\right] = 23.14\left[\frac{1}{7} + \frac{(-.5)^2}{7} + \frac{(-.5)^2}{7}\right] = 23.14(.2143) = 4.9586$$

t critical = 2.10

The desired confidence interval is:

$-2.3575 \pm 2.10\sqrt{4.9586} \Rightarrow -2.3575 \pm 2.10(2.2268) \Rightarrow -2.3575 \pm 4.6762 \Rightarrow (-7.0337, 2.3187)$.

Chapter 16

Exercises 16.1 – 16.7

16.1 Let μ_1 denote the true average fluoride concentration for livestock grazing in the polluted region and μ_2 denote the true average fluoride concentration for livestock grazing in the unpolluted regions.

$H_0: \mu_1 - \mu_2 = 0 \quad H_a: \mu_1 - \mu_2 > 0$

$\alpha = 0.05$

The test statistic is: rank sum for polluted area (sample 1).

Sample	Ordered Data	Rank
2	14.2	1
1	16.8	2
1	17.1	3
2	17.2	4
2	18.3	5
2	18.4	6
1	18.7	7
1	19.7	8
2	20.0	9
1	20.9	10
1	21.3	11
1	23.0	12

Rank sum = (2 + 3 + 7 + 8 + 10 + 11 + 12) = 53

P-value: This is an upper-tail test. With $n_1 = 7$ and $n_2 = 5$, Chapter 16 Appendix Table 1 tells us that the P-value > 0.05.

Since the P-value exceeds α, H_o is not rejected. The data does not support the conclusion that there is a larger average fluoride concentration for the polluted area than for the unpolluted area.

16.3 **a.** Let μ_1 denote the true average ascent time using the lateral gait and μ_2 denote the true average ascent time using the four-beat diagonal gait

$H_0: \mu_1 - \mu_2 = 0 \quad H_a: \mu_1 - \mu_2 \neq 0$

A value for α was not specified in the problem, so a value of 0.05 was chosen for illustration.

The test statistic is: Rank sum for diagonal gait.

Gait	Ordered Data	Rank
D	0.85	1
L	0.86	2
L	1.09	3
D	1.24	4
D	1.27	5
L	1.31	6

L	1.39	7
D	1.45	8
L	1.51	9
L	1.53	10
L	1.64	11
D	1.66	12
D	1.82	13

Rank sum = 1 + 4 + 5 + 8 + 12 + 13 = 43

P-value: This is a two-tail test. With n_1 = 7 and n_2 = 6, Chapter 16 Appendix Table 1 tells us that the P-value > 0.05.

Since the P-value exceeds α, H$_o$ is not rejected. The data does not suggest that there is a difference in mean ascent time for the diagonal and lateral gaits.

b. We can be at least 95% confident (actually 96.2% confident) that the difference in the mean ascent time using lateral gait and the mean ascent time using diagonal gait may be as small as –.43 to as large as 0.3697.

16.5 Let μ_1 denote the true mean number of binges per week for people who use Imipramine

and μ_2 the true mean number of binges per week for people who use a placebo.

$H_0: \mu_1 - \mu_2 = 0 \quad H_a: \mu_1 - \mu_2 < 0$

$\alpha = 0.05$

The test statistic is: Rank sum for the Imipramine group.

Group	Ordered Data	Rank		Group	Ordered Data	Rank
I	1	1.5		P	4	8.5
I	1	1.5		I	5	10
I	2	3.5		P	6	11
I	2	3.5		I	7	12
I	3	6		P	8	13
P	3	6		P	10	14
P	3	6		I	12	15
P	4	8.5		P	15	16

Rank sum = 1.5 + 1.5 + 3.5 + 3.5 + 6 + 10 + 12 + 15 = 53

P-value: This is an lower-tail test. With n_1 = 8 and n_2 = 8, Chapter 16 Appendix Table 1 tells us that the P-value > 0.05.

Since the P-value exceeds α, H_0 is not rejected. The data does not provide enough evidence to suggest that Imipramine is effective in reducing the mean number of binges per week.

16.7 The 95.5% C.I. for $\mu_1 - \mu_2$ is given as (-0.4998, 0.5699). The confidence interval indicates that the mean burning time of oak may be as much as 0. 5699 hours longer than pine; but also that the mean burning time of oak may be as much as 0.4998 hours shorter than pine.

Exercises 16.9 – 16.17

16.9   Let μ_d denote the true mean difference of in peak force on the hand. (Post – Pre).

$H_0: \ \mu_d = 0$ $H_a: \ \mu_d > 0$

Test statistic is the signed-rank sum.

With $n = 6$ and $\alpha = 0.109$, reject H_0 if the signed rank sum is greater than, or equal to 13.

(Critical Value from Chapter 16 Appendix Table 2)

Differences = (Post – Pre) – 10

The differences and signed ranks are:

Differences: 1.5, -2.1, -2.9, -7.1, 7.3, -8.9.

Signed-ranks: 1, -2, -3, -4, 5, -6.

Signed-rank sum: = 6

Since the signed-rank sum of 6 does not exceed the critical value of 13, the null hypothesis is not rejected. There is not sufficient evidence to conclude that the mean post impact force is greater than the mean pre-impact force by more than 10.

16.11   Let μ_d denote the true mean difference in concentration of strontium-90. (nonfat minus 2% fat).

$H_0: \ \mu_d = 0$ $H_a: \ \mu_d < 0$

Test statistic is the signed-rank sum.

With $n = 5$ and $\alpha = 0.05$, reject H_0 if the signed rank sum is less than, or equal to -13.

(Critical Value from Chapter 16 Appendix Table 2)

The differences and signed ranks are:

Differences: -0.7, -4.1, -4.7, -2.8, -1.7.

Signed-ranks: -1, -4, -5, -3, -2.

Signed-rank sum: = -15.

Since the calculated signed-rank sum of -15 is less than the critical value of -13, the null hypothesis is rejected. The data does suggest that the true mean strontium-90 concentration is higher for 2% milk than for nonfat milk.

16.13   **a.** Let μ_d denote the true mean difference in height velocity. (during minus before).

$H_0: \ \mu_d = 0$ $H_a: \ \mu_d > 0$

Test statistic is the signed-rank sum.

With $n = 14$ and $\alpha = 0.05$, reject H_0 if the signed rank sum is greater than, or equal to 53.

(Critical Value from Chapter 16 Appendix Table 2)

The differences and signed ranks are:

Differences: 2.7, 7.6, 2.0, 4.9, 3.5, 7.7, 5.3, 5.3, 4.4, 2.0, 6.4, 4.3, 5.8, 2.6.

Signed-ranks: 4, 13, 1.5, 8, 5, 14, 9.5, 9.5, 7, 1.5, 12, 6, 11, 3.

Signed-rank sum: = 105.

Since the calculated signed-rank sum of 105 exceeds the critical value of 53, the null hypothesis is rejected. it is therefore concluded that the growth hormone therapy is successful in increasing the mean height velocity.

b. The assumption that must be made about the height velocity distributions is that they are identical with respect to shape and spread, so that if they do differ, they differ only with respect to the location of their centers.

16.15 Let μ_d denote the true mean difference in cholesterol synthesis rates. (potato minus rice).

$H_0: \ \mu_d = 0 \qquad H_a: \ \mu_d \neq 0$

Test statistic is the signed-rank sum.

With $n = 8$ and $\alpha = 0.05$, reject H_0 if the signed rank sum is less than or equal to -28, or greater than or equal to 28. (Critical Value from Chapter 16 Appendix Table 2)

The differences and signed ranks are:

Differences: 0.18, -1.24, 0.25, -0.56, 1.01, 0.96, 0.60, 0.16

Signed-ranks: 2, -8, 3, -4, 7, 6, 5, 1.

Signed-rank sum: = 12

Since the calculated value of 12 does not fall into the rejection region, H_0 is not rejected.

The data suggests that there is no difference in the true mean cholesterol synthesis for the two sources of carbohydrates..

16.17 Using μ_d as defined in Exercise 16.13, and the differences computed there, the pairwise averages are:

	2.0	2.0	2.6	2.7	3.5	4.3	4.4	4.9	5.3	5.3	5.8	6.4	7.6	7.7
2.0	2.0	2.0	2.3	2.35	2.75	3.15	3.2	3.45	3.65	3.65	3.9	4.2	4.8	4.85
2.0		2.0	2.3	2.35	2.75	3.15	3.2	3.45	3.65	3.65	3.9	4.2	4.8	4.85
2.6			2.6	2.65	3.05	3.45	3.5	3.75	3.95	3.95	4.2	4.5	5.1	5.15
2.7				2.7	3.1	3.5	3.55	3.8	4.0	4.0	4.25	4.55	5.15	5.20
3.5					3.5	3.9	3.95	4.2	4.4	4.4	4.65	4.95	5.55	5.60
4.3						4.3	4.35	4.6	4.8	4.8	5.05	5.35	5.95	6.0
4.4							4.4	4.65	4.85	4.85	5.1	5.4	6	6.05
4.9								4.9	5.1	5.1	5.35	5.65	6.25	6.3
5.3									5.3	5.3	5.55	5.85	6.45	6.5
5.3										5.3	5.55	5.85	6.45	6.5
5.8											5.8	6.1	6.7	6.75
6.4												6.4	7.0	7.05
7.6													7.6	7.65
7.7														7.7

With $n = 14$, and a confidence level of 90%, $d = 27$. Counting in 27 averages from each end of the ordered pairwise averages yields the confidence interval of (3.65, 5.55). Thus, with 90% confidence, the true mean difference in height velocity is estimated to be between 3.65 and 5.55 units; that is, there is greater growth velocity during the therapy than before the therapy.

Exercises 16.19 – 16.25

16.19 Let μ_1, μ_2, and μ_3 denote the true average importance scores for lower, middle, and upper social class, respectively.

$H_0: \ \mu_1 = \mu_2 = \mu_3$

$H_a:$ at least two of the three μ_i's are different

Test statistic: Kruskal Wallis

Rejection region: The number of df for the chi-squared approximation is k – 1 = 2. For α = 0.05, Chapter 16 Appendix Table 4 gives 5.99 as the critical value. H_0 will be rejected if KW > 5.99.

KW = .17 as given. Since this computed value does not exceed the critical value of 5.99, the null hypothesis is not rejected. The data does not provide enough evidence for concluding that true average importance scores for lower, middle and upper social classes are different.

16.21. Let μ_1, μ_2, μ_3 and μ_4 denote the true mean phosphorous concentration for the four treatments levels for the normal workers, alcoholics with siderblasts and alcoholics without siderblasts, respectively,

H_0: $\mu_1 = \mu_2 = \mu_3 = \mu_4$

H_a: at least two of the four μ_i's are different

Test statistic: Kruskal Wallis

Rejection region: The number of df for the chi-squared approximation is k – 1 = 3. For α = 0.01, Chapter 16 Appendix Table 4 gives 11.34 as the critical value. H_0 will be rejected if KW > 11.34.

Treatment			Ranks			$\bar{r}_i$
1	4	1	2	3	5	3
2	8	7	10	6	9	8
3	11	15	14	12	13	13
4	16	20	19	17	18	18

$$KW = \frac{12}{(20)(21)}\left[5(3-10.5)^2 + 5(8-10.5)^2 + 5(13-10.5)^2 + 5(18-10.5)^2\right] = 17.86$$

Since 17.86 > 11.34, the null hypothesis is rejected. The data strongly suggests that true mean phosphorous concentration is not the same for the four treatments.

16.23 H_0: The mean permeability does not differ for the four treatments.

H_a: The mean permeability does differ for at least two of the four treatments.

Test statistic: Friedman

Rejection region: With α =0.01, and k - 1 = 3, Chapter 16 Appendix Table 4 gives a chi-square value as 11.34. H_0 will be rejected if F_r > 11.34.

	Ranks										
	Subject (blocks)										
Treatment	1	2	3	4	5	6	7	8	9	10	$\bar{r}_i$
1	2	2	2	2	2	2	2	2	2	2	2
2	1	1	1	1	1	1	1	1	1	1	1
3	4	4	4	4	3	4	4	4	4	4	3.9
4	3	3	3	3	4	3	3	3	3	3	3.1

$$F_r = \frac{(12)(10)}{(4)(5)}\left[(2-2.5)^2 + (1-2.5)^2 + (3.9-2.5)^2 + (3.1-2.5)^2\right] = 28.92$$

Since 28.92 > 11.34, the null hypothesis is rejected. The data does provide evidence to suggest that the mean permeability is not the same for at least two of the four treatments.

16.25 H_0 : The mean survival rate does not depend on storage temperature.

H_a : The mean survival rate does depend on storage temperature.

Test statistic: Friedman

Rejection region: With α =0.05, and k - 1 = 3, Chapter 16 Appendix Table 4 gives a chi-square value as 7.82. H_0 will be rejected if $F_r > 7.82$.

| Temperature | \multicolumn{6}{c}{Ranks} |
| | \multicolumn{6}{c}{Storage Time} |
	6	24	48	120	168	$\bar{r}_i$
15.6	3	3	3	3	3	3
21.1	4	4	4	4	4	4
26.7	2	2	2	2	2	2
32.2	1	1	1	1	1	1

$$F_r = \frac{(12)(5)}{(4)(5)}\left[(3-2.5)^2 + (4-2.5)^2 + (2-2.5)^2 + (1-2.5)^2\right] = 15$$

Since 15 > 7.82, the null hypothesis is rejected. The data suggests that the mean survival rate differs for at least two of the different storage temperatures.